On the Economics
of Solar Energy

On the Economics of Solar Energy

The Public-Utility Interface

Stephen L. Feldman
University of Pennsylvania

Robert M Wirtshafter

LexingtonBooks
D.C. Heath and Company
Lexington, Massachusetts
Toronto

Library of Congress Cataloging in Publication Data

Feldman, Stephen L

  On the economics of solar energy.
    1. Solar energy industries—United States. 2. Public utilities—United States. 3. Solar energy—Economic aspects—United States. 1. Wirt-shafter, Robert M., joint author. II. Title
HD9681.U62F44   338.4'762147'0973   79-5442
ISBN 0–669–03449–5

Copyright © 1980 by D.C. Heath and Company.

Second printing, May 1981

Printed simultaneously in Canada.

Printed in United States of America.

International Standard Book Number: 0–669–03449–5

Library of Congress Catalog Card Number: 79-5442

Contents

                     Credits to Consumers, and Utility Financing of
                     Solar Energy Systems   145

                     Observations on Utility Rate Structures and
                        Decentralized Solar Energy Consumers   145
                     Public Service Company of New Mexico at
                        Albuquerque   173
                     Pacific Gas and Electric
                     Los Angeles Department of Water and Power   174
                     Implications for Public Policy   179
                     Utility Financing of Solar Applications and
                        Energy Conservation   180

Chapter 6   **Public Utilities and Solar Energy Development—**
                     Institutional Economic Considerations
                     Roger G. Noll   211

                     Solar Energy and Centralized Utilities   212
                     Alternative Regulatory Policies   213
                     Arguments against Solar Ownership by Utilities   215
                     Problems of Liability and Quality Control   217
                     The Case for Combined Solar Utilities   219
                     Some Solutions to the Policy Problem   225
                     Conclusions   228

Chapter 7   **The Solar-Utility Interface: A Capsule Appraisal**   231

                     References   239

                     About the Authors   257

List of Figures
and Tables

Preface
and Acknowledgments

Any interdisciplinary attempt at modeling systems requires that the investigators be familiar with the language and precepts of their colleagues. Working with architects and engineers for geographers *cum* economists like ourselves has been one of the most exuberant of intellectual experiences. Our growth has been the result of the generosity of others—of not only the authors of the appropriate sections of this text but also those who have passed through our corridors for short but ebullient periods of our enlightenment. The list is long, and we must be forgiven for not mentioning everyone.

Supportive of this effort from the outset have been those who financially and intellectually sponsored us. From the National Science Foundation and U.S. Department of Energy were Laurence Rosenberg and Roger Bezdek, respectively. As president of the Russell Sage Foundation, Aaron Wildavsky was deeply committed to seeing the development of policy analysis by micromodeling. Others have provided support to allow us to refine our techniques and methods and to adopt them to case studies that had policy value: Ron Doctor of the California Energy Commission, Ernst Habicht, Jr., of the Environmental Defense Fund, and T. Louis Austin of Texas Utilities Company were particularly responsive.

A host of students and colleagues aided the effort: Michael Abrash, Michael Berz, John Breese, Carol Carter, Jeremy Coleman, Charles Dickson, Martin Holdrich, Joseph Kohler, David Magid, Thor Olsen, Martha Simon, Brian Straub, Paul Sullivan, Mark Wallenrod, and Paul Winoski. The officials from some fifteen public utilities who volunteered their time to endure with our seemingly endless data requests must be praised by us.

Without the patient advice and contributions of our coauthors, little would have ensured the completion of our interdisciplinary task. They are Bruce Anderson, Charles Cicchetti, William Gillen, Richard Howard, Edward Kahn, Roger Noll, Paul Sullivan, Eliot Wessler, and Douglas Woods.

The manuscript could not have had its timely completion if it were not for the typing skills of Elsie Sullivan and Terry Parise. Neither would it have achieved publication without the constant encouragement of Diane Feldman and Angela Sorrentino.

1

An Introduction to the Problem

The present concern over seeking alternative methods of heating and cooling residential buildings is one problem confronting people in their search for comfort and shelter. The rising relative costs of heating air and hot water and cooling a home apply ever-increasing pressure on household budgets. Technological innovations such as solar energy must be developed if people are to continue to inhabit areas where space conditioning is necessary for comfort. The individual and society must make difficult decisions regarding investments in technology and the continued use of energy sources which are in short supply. The development of criteria for selecting among these technologies is the focus of this book.

In 1975 people in the United States utilized more than 10 quads[1] of energy per year for residential space heating, air conditioning, and hot water. This value represents about 15 percent of the total annual energy use in the United States. Most of this energy involves gas and oil and is therefore subject to supply constraints and foreign intervention. In addition, electricity, which may have a more elastic supply than other energy sources, generally is more expensive for the consumer and has many environmental side effects.

One alternative energy source—solar energy—is being touted by some as being plentiful, inexpensive, and environmentally benevolent. Yet, some applications of solar energy may have high initial investment costs, utilize large quantities of raw materials, and require additional auxiliary energy sources. Solar energy systems usually require auxiliary systems for periods of cloudiness and during the night; as a result, there may be problems in integration with an electricity supply grid.

Despite a number of technicoinstitutional problems with solar heating and cooling technologies, the federal government perceives enough benefits in the potential of these technologies to encourage a program of rapid promotion. Central to this plan is the enactment of incentives for qualified solar installations in the form of tax credits. Unfortunately, this decision by government may not have been based on a systematic analysis which reflects all the concerns for economic efficiency and environmental quality; it suggests, rather, the desire to develop as rapidly as possible technically feasible alternatives to oil and gas. To this end, the government has offered incentives to consumers to increase the cost-effectiveness of particular technologies. This policy fails to include any assessment of solar systems within

1

a wider, total resource management perspective within the domain of decentralized building technologies.

Clearly, a more critical examination of the potential benefits of solar energy is warranted in an effort to determine the consistency of each application of solar energy with an overall program of energy conservation and national economic efficiency. This book examines the potential tradeoffs between energy conservation and solar energy use in the space conditioning of residential buildings and the heating of hot water, as they interface with electric utilities. In addition, this book defines more clearly the interface between solar building design characteristics and the institutional economics of electric utility companies. Also, the analysis is extended to include other technologies and policy options. True benefits of alternative energy applications can properly be assessed only within a systematic examination of the entire supply-and-demand network of the electric utility industry.

This chapter provides background material necessary to a more complete understanding of the role of solar heating and hot water within residential space conditioning. We begin with an analysis of space conditioning and buildings, the historic perspective, and actual physical properties. Second, we discuss the importance of the individual consumer and his decision-making process. Third, we examine the electric utility industry, which is most likely to serve as competitor and supplier of auxiliary energy to solar systems. Finally, we analyze solar energy and energy conservation from a national perspective.

Space Conditioning and Buildings

Human beings, not unlike other creatures, must be able to cope with a dynamic and sometimes harsh environment. Most animals have biological adaptation mechanisms that permit them to change their internal body temperature or to protect themselves from the elements. Human beings have survived because of cultural adaptations to climate. Cultural adaptations including clothing and shelter have enabled human beings to maintain a constant internal body temperature of 37°C despite a broad range of ambient temperatures.

A complex homeostatic system has evolved in humans to control their internal temperature. If a human is too hot, his circulation increases and his pores expand, causing perspiration. Under cold conditions the opposite occurs—the skin and its underlying capillaries contract, and the heart slows down. Clothing becomes necessary to minimize the heat loss. In this manner the internal heat source, the delivery or epidermic system, and the behavioral system work in concert.

Buildings evolved to act similarly to the body itself. Fire provided humans with an internal heat source, and the external structure provided them with a barrier to heat loss. Early architecture was most concerned with the immediate problem of protection from the wind and precipitation and to a lesser extent with providing thermal comfort. Yet an important point is that homes were not developed exclusively because of the need to stay warm. James Marston Fitch (1975) believes that a shelter is a method to limit all external stresses; in effect, a home should offer a completely separate and stress-free environment. Fitch mentions the effects that a building may have on a human body, including thermal, atmospheric, aqueous, luminous, sonic, external world-of-objects, and spatiogravitational. The challenge to the architect is to include all these functions in a building.

Of concern here are those aspects of building design which are capable of reducing the consumption of energy. For example, to some extent, architects have been successful in lowering electric lighting expenses through the proper design and placement of windows. Of more concern to us is the heating and air conditioning aspects of the building.

Historic Heating and Cooling of Buildings

Historically, open fire was the first source of internal heat to a home. But since an open fire was dangerous, fireplaces with a brick enclosure became the dominant mode of heating. A fireplace, though, is an inefficient method of burning wood, because most of the heat is drawn up the chimney. Franklin stoves were developed and became popular because of their higher efficiencies, but also as a result of wood shortages. Each of these methods— fires, fireplaces, and stoves—had another major drawback in that the heat enters the room from a single point source. The desire to maintain a consistent temperature throughout the structure led to the development of central heating systems. These systems generally circulated hot air or hot water (later, cold air or water was used for air conditioning) through the interior of the building. These convection systems were fired by coal and later by oil or gas. A further technological innovation was the use of electric resistance heating. These systems are inexpensive to install and require very little maintenance.

Each of these innovations in heating and cooling technology was brought on by a combination of forces. In most cases, the primary source of fuel was in short supply or required difficult handling problems. Coupled with this was the fact that the new technology offered more consistent performance, usually at a lower price. For an analysis of historic changes in energy supply, technology, and culture, see Lynn White (1962) or James

O'Toole (1976). With each of these innovations, more specialization was necessary. Today maintenance for heating and cooling equipment is performed by specialists, normally a heating, ventilation and air conditioning (HVAC) engineer. This has caused a change in the roles of the architect and architecture in building construction.

The role of the architect used to include building the temperature and humidity control into the plans of the buildings. Early architecture developed homes built out of indigenous materials that often successfully shielded the interior from harsh environments. Victor Olgyay, in *Design with Climate* (1963, pp. 4–5), mentions the igloos of the Eskimos as a brilliant example of a technology well suited to an environment of extreme cold and relatively scarce fuel:

> These low hemispherical shelters deflect the winds and take advantage of the insulating value of the snow that surrounds them. The smooth ice lining which forms on their interior surface is an effective seal against air seepage, and the tunnel exits are oriented away from the prevailing winds to reduce drafts and prevent the escape of warmed air. The heat retention of this type of structure makes it possible to maintain a temperature of 60°F inside when the exterior temperature is − 50°F. Such structures may be heated by a small lamp supplemented by body heat.

Climate Design in Modern Buildings

A consistent theme in much of the recent literature on energy conservation in buildings is the lack of concern by architects for the heating and cooling requirements of buildings. This is clearly demonstrated by the use of nonfunctional windows in many modern office buildings. Recently architects have concentrated their efforts on improving other aspects of residential needs, including aesthetics. Philip Steadman (1975) credits not only the inexpensive cost of heating and cooling systems but also the contribution of functionalism in building design of the 1920s and 1930s. He states that architects such as Corbusier began designing buildings from the inside out. "The result of all these exhortations about 'truth to materials' and to building function, was that the external containing envelopes of buildings became very largely just consequences, almost accidental resultant of their internal organization and material structure" (Steadman 1975, p. 14).

The building industry generally uses two approaches in providing comfort. One relies on the design of the building itself, while the other depends on mechanical systems. Although both approaches are used in all buildings, there is usually an overwhelming dependence on mechanical systems. In fact, buildings are often designed without much thought being given to comfort, and mechanical systems must compensate.

In the field of solar energy, the architecture-based solutions are called *passive*. Passive systems normally do not have mechanical devices, but rather incorporate the potential for solar utilization for space heating within the building design. Passive solar buildings often utilize extensive glass areas on the building's southern exposure in conjunction with large amounts of building thermal mass. The building thermal mass (usually concrete or water storage contained within the building structure) is heated by solar radiation during sunny days and releases stored heat to the interior of the building. One advantage of passive buildings over active buildings is that design considerations can also be utilized to maintain cool temperatures in the summer. Often this is accomplished with shutters on south-facing windows and/or overhanging shades. In this case, the thermal mass may store cool air, which is then released to the interior space. Some present solar energy buildings combine both active and passive elements.

The solutions based on mechanical systems are usually called *active,* since mechanical devices, such as fans and pumps, which use external sources of energy in their operations are required to transport thermal energy to maintain comfort. For solar systems, *active* refers to methods which require solar collectors on the building (or next to it) and solar thermal-energy storage as a subsystem tied into the building's mechanical system. The solar collectors may heat air, which then heats either the building or the heat storage container for later use. Or they may heat a liquid, which then heats the building or is stored for later use.

The Building Load

The building utilizes energy for heating or cooling because of the required maintenance of an internal building temperature different from that of the external environment. Three principal aspects of a building affect the magnitude of the building load: the building envelope, the internal loads, and external environment. First, the building-envelope characteristics—the walls, ceiling, insulation, windows, and doors—provide a barrier between inside and outside. The amount of conduction and infiltration through this envelope determines the thermal effectiveness of the structure. Second, internal loads are generated by occupants, appliances, HVAC equipment, and so on. Third, the external environment impacts on the building envelope and the internal heat generation.

For a simple estimate of building load, these factors are combined to depict a homogeneous box in which seasonal energy consumption Q is crudely calculated by using a steady-state UA value to represent the building envelope, along with $Qgen$ and degree-day values for internal heat generation and external temperature:

$$Q = U \cdot A \cdot DD \cdot 24 \, \text{h/day} - Qgen$$

where Q = building energy consumption (Btu)

U = composite coefficient of heat transmittance of the house, the reciprocal of the individual heat resistances [Btu/(h $\cdot$ ft^2 $\cdot$ °F)]

A = surface area of the house (ft^2)

DD = degree-days (°F)—number of heating degrees per season calculated by 65°F minus ambient temperature

$Qgen$ = internal heat generation (Btu)

In these simplified building models the building heat load for the residence in any one period is equated to building heat loss. In treating building cooling loads, the energy flows are reversed, and degree-days are calculated as ambient (or external) temperature minus 75°F.

In actuality, the interaction among the different aspects of the building is not as independent or static as this simple model implies. The external environment affects the building envelope such that the temperature difference determines the amount of conduction through the barrier. Wind alters the heat-loss value by drawing the heat off the building at varying rates, and wind also affects the internal temperature through infiltration. Radiation on the building surface can effectively raise the surface temperature and cause a heat flow from the exterior to interior despite low ambient temperatures.[2]

These flows of energy via conduction, convection, and radiation vary with respect to each surface of the building. More importantly, these flows are not static, so that heat gains in any one time frame are not equivalent to heat losses. Conduction, convection, and radiation flows interact dynamically and at different rates with the building elements affecting heat transfer. This is particularly true when cooling loads are considered, where humidity as well as temperature is important.

Building Internal Gains. Let us begin with the most important aspect of the building load—the human occupants. The home operates to suit their needs. The building internal temperature is controlled to meet their comfort demands. Each human occupant generates about 500 Btu/h in heat energy within the residence in addition to the heat expelled by operating appliances. The nature of individual behavior creates an extremely difficult problem in accounting for these actions within a building load model. [For an example of the difficulties of monitoring buildings and building internal gains, see Socolow et al. (1978).] Yet, individual behavior may account for the largest variation in loads between buildings, and alteration of behavior may prove to be the simplest method of lowering building energy consumption.

Building Design. The importance of the building envelope in determining building load has been established. The actual construction of the walls and ceilings is of prime concern. The type of materials, the construction practices, the design, and the orientation all affect building loads. Building materials resist energy transfer by conduction at different rates. Well-constructed buildings can drastically lower building loads by reducing infiltration. Building design and orientation play the key role not only in lowering heat loss, but also by providing the potential for energy gains. This is particularly true of the use of fenestration (the arrangement and design of windows) and building thermal mass.

Passive Solar Techniques

The integration of insulation, fenestration, thermal mass, and orientation into building design to increase the solar energy potential of the structure is called the *passive solar technique.* Balcomb (1976) divides passive systems into five generic types: direct-gain, mass-storage walls, mass-storage roofs/ceilings, convective loops, and attached sun spaces. Direct-gain systems describe the combination of fenestration and thermal mass highlighted above where solar radiation enters the room through south-facing windows and is absorbed directly into the high mass in the room. Storage walls of concrete or water are often placed in the direct path of the solar radiation to more effectively collect and store the energy. The best developments of this procedure, called Michel-Trombewall after the developer, were performed at the Centre National de la Recherche Scientifique.[3] Another technique incorporates a wall of 55-gal drums painted black and filled with water.[4] Roof ponds or mass storage roofs have been found to be particularly effective in the Southwest, where Harold Hay has developed Sky Therm, in which water bags stored in the roof constitute an ample amount of thermal mass.[5] The ingenuity of this design lies in the movable insulation above the water bags. In the winter the mode of operation is to open the insulation during the day to collect energy and then close it at night to allow the heat to flow into the living area. In the summer the process is reversed, which takes advantage of the radiational cooling effect of cool summer nights.

Some systems of solar radiation collection have been developed which transport hot air or water away from the collector to the rest of the home by thermocirculation. These natural convection systems eliminate overheating problems by circulating the collected energy through the building and drawing the coldest air back to the collector. Finally, a fifth category of passive techniques includes sun spaces such as attached greenhouses.

Many of the passive buildings presently being constructed utilize more than one of these five generic types within the design. Because of the complexities and interaction within the building design, clear definitions of passive systems have not been established. [See Passive Solar Heating and Cooling Conference (1976, p. 167).] A problem is the distinction presently made by most legislation that permits tax savings for active systems only. Another constraint to the development of passive systems is the inability to adequately model the performance of these systems.

Meeting Building Loads—HVAC Equipment

While proper building design can significantly lower building loads, in severe climates it is not always sufficient to maintain comfortable internal temperatures. Heating and cooling are supplemented by the use of mechanical HVAC equipment. Because the discussion of all available technologies is beyond the scope of this book we focus on one potential technology—active solar energy systems. Because other technologies, including oil and gas furnaces, electric heat pumps, and electric radiative heating, all have the potential to lower space conditioning costs, these technologies can and should be evaluated by using a similar methodology.

One point that does warrant discussion is the method of sizing HVAC equipment. The HVAC equipment is selected to ensure comfort maintenance at "design conditions," the worst possible weather conditions. The sizing at design conditions stems from the fact that HVAC equipment must maintain comfort levels but also protect the home from damage, for instance, from freezing pipes. The dependence on the HVAC equipment for meeting design conditions has particular importance in the discussion of new energy-saving devices such as solar energy systems or heat pumps. Because these systems generally are sized to meet only average conditions, some protection against design conditions is still required.

Recently in some cases, heating and cooling bills have become a staggering part of a homeowner's expenses, sometimes higher than the mortgage payment. High operating costs for presently used space-conditioning systems have led individuals to search for alternatives such as solar energy. The opportunity for energy savings by the design of buildings may also exist and should therefore be pursued in conjunction with improved energy delivery systems. The optimization of the heating and cooling requirements of a building should be incorporated into the overall complex functions of the building. In some cases, treating the space-conditioning aspects of the building as separate from the design has led to inefficient use of resources. In addition, the lack of concern for overall building performance and design has led to significant difficulties for electric utilities that must serve these buildings. Three principals are primarily responsible for this interface: the homeowner who builds the home and uses the electricity, the electric utility that provides the electricity and sets the electric rates, and the government that has regulatory responsibility for the entire process.

The Homeowner—His Decision Process

Conceptually, the building process centers on the individual that eventually buys the home, though often this individual does not initiate the building process. In the end, however, decisions about the type of space-conditioning system are made by the homeowner. It is necessary to determine the importance of this decision with respect to the house-buying process.

The variety of functions that a residence must provide has been described. To the individual, the combination of these characteristics which blend into the desired home is a complex relationship between sociological, psychological, philosophical, aesthetic, and economic factors. Because the purchase of a residence is one of the largest investments most people make, economic concerns normally weigh heavily in any final decision. Of concern to this book is the extent and nature of the economic decision-making process with regard to space conditioning and domestic hot water.

Economic Assumptions—Perfect Competition

At first, let us assume that the individual consumer makes his purchase decisions within a set of perfect market conditions: without monopoly power, with free information and movement, without increasing returns to scale, and without externalities. Given this perfect-competition marketplace, the individual maximizes his utility (or value) subject to his budget constraint by equating the marginal benefits of the last unit of a good purchased to the marginal cost of that good. The individual will purchase just enough space conditioning so that the next unit of space conditioning obtains the same ratio of marginal benefits to marginal costs as the next unit of decoration, plumbing, and food for the family. [For an analysis of economic assumptions of perfect competition and consumer price theory, see Samuelson (1947).]

Some Qualifications of Perfect Competition

One problem is the assumption that the individual possesses perfect knowledge, so that he is capable of determining the set of options which maximize his benefits from the infinite set of conceivable possibilities. Normally, the purchaser's realm of knowledge does not extend beyond the few different building types presented by the builder. Nor is the individual perfectly cognizant of his own utility function so as to select the optimum configuration. Generally a certain level of space conditioning is requested or required through building codes, and the homeowner bases his decision on finding the option that obtains these performance levels at the lowest cost.

A second assumption that does not apply to this decision process is that of free movement of goods and services. This assumption permits the movement of capital from one investment to another without cost. Unfortunately, much of the investment capital in a home is a fixed, first-time investment. A decision by a consumer to change space-conditioning systems may require significant additional expense, as is evident by the greater cost of retrofitting solar heating systems in existing buildings versus installing solar heating system in new construction.

The fact that the space-conditioning system must be fixed at the time of construction alters the decision process of the individual. Some long-range decisions are in order. The decision process now must account for predictions of future cost and the value of future benefits.

For convenience, it is best to divide the costs of a space-conditioning system into two categories: capital costs and variable costs. Capital costs are those equipment purchases which are fixed at construction time. Variable costs include the cost of fuel and maintenance to operate the space-conditioning equipment. Some space-conditioning options such as electric resistance heating have low initial investment costs, but require large operating expenditures. Investments in solar energy increase the initial outlay of capital, but lower running costs. A set of space-conditioning options is optimized when the total cost with respect to total benefits is maximized over the total lifetime of the equipment. (Some adjustment from lifetime of equipment to lifetime of occupancy is warranted if the equipment does not retain resale potential. For example, the average duration of occupancy within the United States is about five years and therefore creates a great need for the marketing of solar homes as a better purchase despite higher initial cost.)

The individual optimization of this future stream of costs varies with individual perceptions of the future. As evidenced by the variability of personal savings among individuals, these perceptions are diverse. A recent survey has indicated that initial costs pose the greatest barrier to rapid commercialization of solar energy, despite the existence of future savings (Opinion Research Corp. 1976).

Within our set of market assumptions there are several other qualifications that affect the consumer's final decision. The decision to build a home does not always rest with the eventual buyer. The building industry dictates many constraints which limit both the options available to the homeowner and the commercial availability of many beneficial building designs.

The Impact of the Building Industry on the Consumer

Unlike most major industries, the building industry is diverse and uncon-glomerated. The number of principals involved in the construction of a building, the general competitive nature of the industry, and the cost of labor make it difficult for the individual contractor to be anything but conservative. As Schoen, Hirshberg, and Weingart (1975, p. 51) state,

> If new technologies for buildings require installation techniques that in-crease the construction time (and therefore construction costs), it is un-likely that they will be accepted rapidly by the industry even if they have other economic advantages. These characteristics tend to make the industry very first cost sensitive, since an easy way to reduce the risk intro-duced by high finance charges is to reduce initial capital requirements. New energy devices which have lower operating costs but higher first costs than do other energy systems may meet resistance.

The responsibility for adoption of new techniques lies with the home-owner who must overcome these barriers and insist on new energy devices. Most homeowners, however, are ill equipped to make decisions involving energy-conserving options. They lack the knowledge to make decisions and are not willing to accept the risks involved in the early adoption of a new technique. Until now the major decision criterion for space-conditioning options has been price. Certain qualifications may restrict the individual consumer's assessment capabilities, but in general the decision is economic within the limited knowledge available. There may exist a segment of the population for whom criteria other than financial ones are important. Recent research by Sawyer and Feldman (1978) of adopters of solar energy systems has revealed that environmental concerns, the desire to tinker, and snob appeal all have led consumers to build or install solar energy devices. It is interesting to note that economic considerations play a large role in the decision process of many of those surveyed. In addition, it is probable that later adopters of solar energy will be even less concerned with these noneco-nomic criteria.

The analysis of consumer cost-effectiveness presented here designates the optimal investment strategies available to consumers. The individual decision process will deviate from the optimal path as a result of inefficien-cies and barriers. An even more critical discrepancy may exist between the consumer decision process and maximization of social benefits. Some costs of energy supply may be external to the individual decision process. One clear example of this lies in the interface between buildings and electric utilities.

The Solar Building/Electric-Utility Interface

Despite a strong desire by many solar energy adopters to be totally self-sufficient with regard to energy, the cost of such an approach normally is excessive. Periods of extreme cold or cloudiness require large collectors and storage systems whose costs may not be justified. For this reason most solar energy systems utilize a backup energy source. Typically this backup source is electricity, largely because it has a low cost of capital equipment to the consumer and is easy to integrate into the solar system. [Over one-third of the 120 buildings listed in Shurcliff (1978) rely on electricity for backup to some degree. Some of these homes also utilize wood and several utilize electricity only during off-peak hours.]

The electric utilities must provide for this electricity demand by maintaining sufficient additional generation capacity to meet this intermittent need. Many utility managers fear that solar energy collection systems will fail at times coincident with peak demand of the utilities system. This will require the electric utility to provide generation capacity equivalent to an all-electric home. Because the solar building's electrical demand is intermittent, the present electricity rate will not return adequate revenue to cover this generation capacity requirement. This situation led one report to conclude that "conventional electric utility systems and most solar energy systems represent a poor technological match" (Asbury and Mueller 1977).

Much of Asbury and Mueller's conclusions are based on the premise that conventional solar energy systems will always have capacity (kilowatt) requirements similar to those of conventional buildings with electric resistance heating. Research by Feldman and Anderson (1976a) and Feldman et al. (1979) has demonstrated that the kilowatt capacity requirement of the solar energy building is not necessarily equivalent to that of a similar all-electric building. Building characteristics, utility characteristics, and weather *together* determine the extent of the impact. Given proper incentives, solar energy investment may provide load management opportunities economically superior to present electric utility load management techniques.

Many researchers have postulated that solar heating and cooling of buildings (SHACOB) will impact negatively on electric utilities which supply backup energy. Feldman and Anderson (1976a) in some cases found that solar buildings adversely affect the load factor and cause a net revenue loss for the utility. They also found that solar cooling in summer-peaking utilities placed very high demands at time of system peak. This is due to the poor coefficient of performance for a solar absorption chiller as opposed to conventional compression equipment. Lorsch (1977) found that SHACOB worsened load factors and exacerbated peak demands for two Pennsylvania electric utilities. Asbury and Mueller (1977) postulate that SHACOB can only displace off-peak fuel and therefore must compete with not only alternative fuels like natural gas and electricity, but also thermal energy storage (TES) applications "fueled" by inexpensive off-peak electricity. The Puerto Rico Water Resources Authority (Llavina 1976), on the other hand, has concluded that solar hot-water commercialization in Puerto Rico will have beneficial effects on the utility's load factor and peaking requirements.

The importance of this interface to the cost-effectiveness of particular technical building options lies in the role that electricity rates play in determining cost-effectiveness. If particular buildings impose costs on electric utilities and do not provide commensurate revenues, then new rates may be justified. The building owner who does not anticipate such rate changes will be incorrectly optimizing his solar system.

Societal benefits of particular building options must also be assessed with regard to the electric utility interface to ensure the development of public policy consistent with maximum societal welfare.

Recent History of the Electric-Utility Industry

Until 1970 the unit cost of kilowatt hours (kWh) supplied decreased as supply increased because fixed costs were spread over more kilowatt hours and new plants were successively more efficient. This is the condition of decreasing long-run marginal costs. This condition probably no longer holds for the electric utility industry, since the rate of inflation for utility costs outweighs cost reductions as a function of advances in technology.[6]

During the period prior to 1970 when marginal costs were decreasing, the utility industry made little effort besides improving generation efficiency to increase all aspects of their production efficiency. Costs were low and so growth in electric loads seemed to benefit everyone—the utility as well as consumers, who experience real price decreases for electricity. Growth benefited the utility and its investors by increasing the capital expenditures on which profits were based, since regulatory agencies allowed the electric utility industries to earn revenues sufficient to cover all costs including a rate of return on capital investment. This created an atmosphere in which capital expenditures was favored and few incentives existed for cost minimization. Utilities implemented a rate structure that promoted the consumption of electricity. Rate schedules often employed the form of "declining blocks" in which the unit price per kilowatt hour decreased as consumption increased. As a result, the rate of growth in electricity consumption rose by 7 percent per year [Edison Electric Institute (1977)] during this period.

Between 1973 and 1977, the rise in the residential electricity price index increased by 16 points more than the rise in the consumer price index. During this period the price of an average kilowatt hour of residential electricity increased by 1.4 cents. A number of causes for this rapid increase are suggested. The price of fuel used by the electric utilities increased drastically over this period. Fuel costs alone, however, are not fully responsible for the increase in costs. Other costs, including labor and the cost of installing new capacity, have also risen. Adding to the cost are the recent requirements for air and water pollution controls which the utilities must install. In addition, clean-air standards required some utilities to discontinue use of fuels from highly polluting but less expensive sources.

Some Present Problems for Electric Utilities

This rapid increase in costs has created a series of problems now being confronted by the electric utility industry, including meeting growth, forecasting demand, and improving operational efficiency.

Meeting Growth. A major problem facing the electric utility industry is its difficulty in raising capital. Capital is needed to expand capacity to accommodate growth in load. Investor-owned utilities raise capital in two ways: by equity financing and debt financing (Finder 1977). During the time that utilities were experiencing decreasing costs, profits were high. Now because of macroeconomic conditions capital is less available, and the utilities must avail themselves of capital markets in which they must compete with alternative investments. The capital-intensive nature of electric utilities hampers their procurement of needed capital because of the vast quantities of capital required and regulatory restrictions on the rate of return on capital expenditure.

As a result of the capital constraints, electric utilities have experienced and will experience difficulty in increasing production. As growth in electricity demand continues, new generation facilities are required. In addition, new and more efficient facilities are required to replace older units.

A further problem restricting a utility's capacity expansion is the time lag between facility planning and on-line operation. The complicated technical nature of modern electric generation facilities requires considerable time for planning and construction. Further exacerbating the long lag time are the environmental impact provisions, facility siting disputes, and other regulatory requirements which delay construction. Thus the construction delays and additional costs force utility companies to anticipate and accurately plan for increases in demand far into the future.

Forecasting Demand. The problem of forecasting changes in demand for electricity is much more difficult and far more critical than it used to be. There is now greater regulatory scrutiny involved in individual utility forecasts. The forecasting problem is made more difficult by the uncertainty in assessing the impacts of price changes on demand. Rate increases and rate structure changes will alter future demands. This can be demonstrated by the halt or decrease in growth of demand experienced by electric utilities following large electricity price increases in the period following the oil embargo.

Certain forces are moving in the direction of increasing demands, including increased saturation of existing appliances such as air conditioning, use of new electric appliances, growth in commerce and industry, and the substitution of electricity for fuels with more elastic supplies. Counterbalancing these trends will be conservation efforts, the development of alternative fuels, and the introduction of competing technologies such as solar energy. Policies such as mandated conservation through prescriptive building codes will also alter future consumption.

A final difficulty in accurately forecasting future demand lies in the role which weather plays. The demand for electricity is sensitive to weather, and so those factors which contribute to future growth and which are weather sensitive take on added importance. Of particular concern to electric utilities are the market penetration of electric resistance heating, air conditioning, and the commercialization of solar energy and other new space-conditioning technologies.

Electric utilities are charged with always meeting demand (in whatever quantity) at the instant at which it is demanded. Therefore, the utility must maintain sufficient capacity to meet the largest demand expected. In addition, the utility must maintain extra capacity above the expected demand, which serves as backup in the event of equipment failure and performs during scheduled maintenance. This extra capacity is called the *reserve margin*. The determination of the reserve margin is a function of reliability criteria chosen by the utility. A typical criterion is no more than two failures (inability to meet load) or 1.5 days in 10 years. In spite of the additional costs of maintaining the reserve capacity, such a high reliability criterion is established because of the even higher cost (economic and social) of power blackouts and service curtailments.[8]

Poor Operating Efficiency. Since the amount of capacity maintained by the utility is normally determined by the annual system peak, during most of the year there is much capacity which is not required to meet average demands. For example, a summer-peaking utility may experience average demands during nonsummer periods that are only 50 percent of the annual system peak. In general, 50 percent of the capacity is idle during these nonsummer periods. This represents considerable inefficiency in that capital resources are not being fully utilized.

Another concern is the diurnal variations in demand. For most utilities, load normally reaches a peak between midafternoon and early evening and is lowest in the early morning. The diurnal variations in load are significant because combinations of generating capacity must be used to meet the varying demand levels during the course of the day.

The utility attempts to utilize as much as possible its larger, more efficient plants for the base-level load. Large fossil-fuel and/or nuclear plants are preferred since they utilize fuel most efficiently. These baseload plants, however, cannot be frequently started up and shut down and therefore run continuously, except for maintenance. Since the utility lacks the technological ability to store electricity, variations in load cannot be handled by these large facilities.

In order to meet these daily peaks, smaller and less efficient generation plants, which are capable of quick start-up, must be utilized. These peaking plants normally use expensive fuels and burn them inefficiently so that incremental costs per kilowatt hour are as high as two to ten times those of the baseload plants.[9]

To summarize the electric utility problems to this point, the availability and high cost of capital have constrained the electric utility's ability to construct new facilities, despite growth in demand. (Since 1973–1974 a slowing growth in demand for capacity as a result of price increases has occurred.) Variations in loads create inefficiencies that cause underutilization of generation facilities and the use of expensive fuels. An additional constraint on the operation of electric utilities is their inability to procure fuel to run their generation facilities. This is particularly true for utilities which have a high proportion of oil or natural-gas generation facilities. For those electric utilities which have a large dependence on hydroelectric capacity, the availability of water (as fuel) may also be a problem. (The capacity of hydropower may, in fact, be downrated as a result of drought experience.)

A partial measure of the efficiency of the use of generating facilities is the *load factor,* which is normally defined as the average annual level of load in kilowatts divided by the system annual peak load in kilowatts. For many electric utilities the load factor is in a range of 0.5 to 0.6 and has remained fairly constant over time. These low load factors indicate that utilities have considerable investment in generating facilities which are used intermittently.

Improving Operating Efficiency and Load Factors. Utility companies have made extensive efforts to find methods of improving load factors while maintaining current standards for loss-of-load probability. Several techniques under the general umbrella of load management have been established by the utilities to make utilization of generation facilities more efficient. The success of several European load management programs is evident by daily load factors as high as 0.95 (Systems Controls 1977, p. 10).

Until recently, U.S. electric utilities had few incentives to improve load factors and minimize costs. As stated previously, the utilities' profits were based on a rate of return on capital investment. In a situation of increasing marginal cost, the utility's ability to maintain profits is reduced as production increases. Utilities are continually requesting price increases to cover real cost increases and expected future cost increases. In general, the regulatory process has been slow in responding to rate-increase requests, which has further eroded the revenue base of the utility. Since the period of large cost increases, utilities appear to be less eager to invest capital in new generating facilities. In response to growing demand, the industry is investigating the potential for load management to ease their capital constraints and at the same time provide valued service.

One way in which the utilities increase their operational efficiency is through a technique known as economic dispatch. Through the use of computers, instantaneous decisions as to how to meet demand are made with the intent of minimizing fuel costs or, in isolated instances, meeting pollution standards and thus maximizing electricity operations. Economic dispatch is based on "system lambda," which is the ranking of the cost efficiency of available generating plants. At the top of the list would be large baseload plants followed by increasingly less efficient plants.

Further optimization of operations can be effected by regionalization of generation facilities, called *power pooling*. Power pooling serves the purpose of spreading individual utility load variations over a larger climatic region. It also enables the individual utility to lower its own reserve margin because of a more extensive pool of reserve.

In Europe utility companies have improved their load factors by shifting some of the peak load to off-peak periods and by lowering the peak demand. Load-shedding techniques allow the utility to reduce some of the peak load by rotating outages to prearranged appliances. Control of these outages is activated by use of decentralized line switches or centralized ripple control.

The filling of demand troughs by shifting demand away from peak demands also improves the load factor. Most of the problems of intermittent demand would be minimized if the utilities possessed inexpensive techniques for centrally storing electric power. The available off-peak electricity could charge storage facilities that would increase peak generation capabilities. Present state-of-the-art storage devices are extremely expensive. The primary method utilized is pumped storage—pumping water uphill during off-peak and releasing the water during peak periods.[10]

Demand-Related Techniques for Load Management. Much of the European success in load management has resulted from changes in pricing policy to encourage shifting of peak demand. In some cases, tariffs designed to promote off-peak usage have successfully induced customers to alter consumption habits and even to purchase hardware to curtail peak usage.[11] The nature of these tariffs varies from country to country, but under these rates those customers who need additional generation will pay for that capacity. Customers may decide that the added expense of electricity at the time of peak demand may make it more beneficial to voluntarily shift to an off-peak period.

Increasing interest has been shown by electric utilities and public utility commissions in altering electricity tariff designs so that electricity prices better reflect the actual cost of production. Under pure competition, the price of a commodity is set at the marginal cost of producing the next unit of that commodity. For regulated monopolies such as the electric utilities, a whole body of literature has proposed setting rates at the marginal cost. Generally the principle of marginal-cost pricing is that, to improve efficiency in electricity utilization, the rates should be established so that the major margins in costs between services are reflected in the tariffs. Criteria other than cost, such as promotional effects or elasticity differences previously used to determine rates, should be deemphasized. The electric utility industry and the federal government are heavily committed to tariff reform. Currently a large number of demonstration projects are supported by the Federal Energy Regulatory Administration and designed to test the benefits and costs associated with marginal-cost pricing and other experimental tariff designs.

Considerable confusion and lack of knowledge about the thermal performance of buildings hinder both the solar energy industry and the utility industry in their efforts to deal with the solar building/utility interface. Buildings are, in fact, thermodynamically complex; there are numerous methods of using solar energy. In some cases solar buildings may actually have a *positive* impact on a utility's financial picture; and there are numerous peak-mitigating solutions for dealing with the solar building/utility interface.

Qualitative Impacts of Buildings on Electric Utilities[a]

Solar building innovations can be divided into solutions based on architecture and on mechanical systems:

1. Solutions based on architecture.
    a. Energy conservation
    b. Collection/rejection

---

[a]This section was written by Bruce Anderson and Paul Sullivan.

    c. Thermal mass
    d. Combined collection and storage
2. Solutions based on mechanical systems
    a. Energy conservation
    b. Systems without electric or gas backup
    c. Systems with gas or electric backup (collectors and systems design)

Each solution is discussed and evaluated in two matrices: (1) the potential impact on the utility energy/demand balance and (2) the matrix of decision-making parameters. The first matrix evaluates the solutions with regard to the manner in which it interfaces with utilities. As has already been described, in most utility systems, total revenues are carefully balanced against total costs of electrical energy consumption and generating capacity.

It is generally believed that solar buildings use less energy than conventional buildings but maintain demand during peak hours in a manner similar to a conventional building. However, this is not necessarily the case. In fact, there are numerous solar building design solutions which can either maintain the energy/demand balance or reduce utility peak loads by a greater factor than the corresponding reduction in energy consumption.

The second matrix evaluates each potential solution with regard to a number of decision-making parameters: site; climate; building size, type, and configuration; utility load profile; use patterns; and comfort. No single solution is cost-efficient on every building site, in every climate, for every building, for every utility, or for every use pattern.

Solutions Based on Architecture

With the rising costs of energy and mechanical devices, good building design as a method for maintaining human comfort becomes increasingly important. Not only is good design the best first step in providing comfort, but also it greatly simplifies any intended mechanical system. In solar energy applications, the building must be a collector of solar heat (and/or a collector of cool and the rejector of heat), and it must store the heat (or cool). These two activities (collector/rejection and thermal storage) are discussed first separately where they are two different and distinct processes and then together as designs in which they occur simultaneously.

Energy Conservation. It should go without saying that energy conservation is the best first step in solving energy problems. This is usually, but not necessarily, true for the solar building/utility interface. (As a comparison, a recent study of the Texas Utilities Company, Feldman has shown decreasing building load factors with improvements in energy conservation in a number of cases.) Most energy-conservation design methods maintain the energy/demand balance of utility companies; that is, energy consumption and demand are reduced by corresponding amounts.

There are many energy-conservation features, however, which can have a favorable impact on the balance. For example, air infiltration increases with an increase in difference between indoor and outdoor temperatures. Therefore, weather stripping and other features to reduce air infiltration can significantly reduce overall energy consumption by a building yet will decrease its peak demand by an even greater fraction. Incorporating thermal mass into exterior walls may increase the wall's resistance to heat loss only slightly, yet it may reduce peak cooling loads considerably by dampening the effects of solar radiation on the building.

Collection/Rejection. Glass (and other transparent membranes) acts as a solar collector. The proper use of glass in building design can provide a significant amount of solar heat gain, thereby reducing the heating load on a conventional backup system and the accompanying mechanical solar heating system. Just as collector design (of the "hardware" variety) can affect the interface with utilities, so can window design. Improper use of windows, for example, can reduce the heating energy consumption of the building because of solar heat gain but, because of the possible increased heat loss during peak conditions, can increase the peak demand on the utility.

There are many decisions to be made about the proper use of glass and other transparent membranes. The type of membrane, the square footage, the orientation, the tilt, the shading coefficient, and the number of layers must be considered. Heat loss must be reduced through the membrane when it is not acting as a collector of heat. While extended use of glass can have a dramatic effect on building design, the architectural profession is in its infancy with regard to understanding how to use glass to reduce energy consumption.

Solar heat must be admitted into a building during the heating season and, conversely, must be prevented from entering during the cooling season. The heat which does penetrate the building (in addition to that which is produced inside the building) must be rejected. Proper glass design is an important factor in reducing undesirable solar heat gain. Natural ventilation and nocturnal radiational cooling are the two primary methods of natural cooling which can be enhanced by proper building design. Nocturnal radiational cooling is the rejection of heat from building to the cold night sky. Just as a solar collector absorbs sunshine, so also can a heat radiator reject heat. Natural ventilation has tended to be an underrated concept in natural cooling. Proper building design can greatly enhance the potential of natural ventilation as a means of avoiding the use of mechanical energy while still maintaining comfort.

Thermal Mass. Once heat or cool has entered a building, the excess amount over and above what is needed for comfort must be stored. Storage methods are similar for both heating and cooling. Thermal storage or building mass in both active systems and architecture-based systems is the focus of attention as a vehicle for achieving peak-mitigating design. For example, a building with sufficient mass might be able to go through late afternoon and early evening peak periods without requiring auxiliary energy. With sufficient thermal mass and good insulating characteristics, and with some tolerance on the part of the building occupants to uncomfortable temperatures, the building could be brought to comfort level prior to the peak period, at which point the auxiliary energy might be disconnected until the peak is over (say, two to five hours).

There are numerous design variables in making decisions to use thermal mass: the type of mass, for example, concrete, brick, stone, or even water; the quantity; the location with respect to the source of thermal energy and to the space to be heated or cooled; the integration into the construction system; and the method and control of heat transfer into and out of storage. Some considerations in the choice among these variables include the method of interface with the backup system, the building's electrical energy demand profile, the utility load profile, and the acceptable change in indoor temperature during a given period. For example, for schools in the service territory of a utility which is winter-peaking, classes may end before the peak period begins (say around 4:00 p.m.). Disconnecting the electrical energy supply to the mechanical systems and letting the mass of the building carry the temperatures through the peak period of between two and five hours duration might be acceptable since occupancy would be at relatively low levels.

For a summer-peaking utility service territory, building mass could help delay the time at which the building experiences its peak cooling demand. Not only might thermal mass delay the time, but also it might decrease the size of the peak. In delaying the time, thermal mass may enable the building to experience its peak well past the peak period of the utility company (say, early evening instead of 3 p.m.). In effect, this could be performed through large time lapses in cycling air-conditioning systems.

Combined Collection and Storage. In many architecture-based solutions, transparent membranes combine with thermal mass to form a design solution which simultaneously collects and stores heat (or cool). Very often the device also serves as an exterior surface of the building, thereby being labeled an *intrinsic* solution.

As previously mentioned, the Skytherm concept of heating and cooling combines heat collection with heat storage during the winter and heat rejection with cool storage during the summer. The most important application of this concept, developed by Harold Hay, is in Atascadero, California, on a house which is 100 percent solar-heated and cooled by the system. The system consists of roof thermoponds in combination with movable insulation. During sunny winter days, the insulation is removed from the roof ponds, exposing the water to the sun. At night the insulation covers the roof ponds, trapping heat which in turn radiates through the ceiling below, heating the space. During the summer, the insulation protects the water from being heated during the day and is removed at night to allow the ponds of water to radiate to the cold night sky. The water in turn cools the house by the natural movement of heat through the cooling to the space below.

A second intrinsic solution makes use of a heat storage wall in combination with a transparent cover. It has been used on hundreds of buildings throughout the world. The sun penetrates the glass and heats the thermal mass, whether it be slate, concrete (developed by Felix Trombe/Jacques Michel in Odeillo, France), or water (in Steve Baer's solution contained in 55-gal drums stacked vertically). [See Anderson (1977).] In cold climates or when increased efficiency is desired, exterior movable insulation covers the transparent membrane at night to reduce the heat loss from the thermal mass to the outside.

Finally, the most simple passive design makes use of large surface areas of glass in combination with sufficient thermal mass inside the building to contain the heat (or cool), thereby maintaining comfort and eliminating wide fluctuations in temperature. In many designs, thermal mass is considered *remote* and may require the use of a circulating fan to transport overheated room air through the thermal mass, such as gravel, to be stored for use when the sun is not shining. Very often, movable insulation covers the glass at night to reduce heat loss.

Solutions Based on Mechanical Systems

Although good building design can significantly reduce dependence on mechanical equipment in providing comfort, most buildings still require complete mechanical systems. With regard to mitigating utility peak loads, there are three areas of approach: energy conservation, elimination of utility backup, and wise use of utility backup.

Energy Conservation. Energy conservation in the design of mechanical systems is particularly applicable to HVAC systems in large buildings. Many systems, such as heat pumps, reduce electric energy consumption by a greater factor than the corresponding reduction in peak demand. In fact, in cold climates heat pumps require a full-sized backup system even though their use during the entire season may result in a total energy savings approximating that of solar energy systems. Thus, there is already a precedent for building comfort systems which significantly upset the energy/demand balance of the utility company compared with electric resistance heating.

Systems without Electric or Gas Backup. The best way of mitigating peak utility demand is to totally avoid the use of gas or electricity. There are two ways of achieving this: use no backup at all (100 percent solar or natural heating and/or cooling), or use a backup source of energy other than gas or electricity. Good building design, high internal heat loads, favorable (mild) climates, wide inside temperature fluctuation, and 100 percent solar heating are the primary methods of satisfying comfort requirements without a backup heating system. Although nocturnal radiation and natural ventilation were discussed under architecture-based solutions, both methods can be enhanced through the use of mechanical systems.

Solar-actuated cooling can take several forms. The heat from solar energy can be used to drive absorption refrigeration equipment or Rankine-cycle compression cooling equipment or to regenerate dessicants for dehumidification. Systems can be designed to provide 100 percent solar cooling.

Coal, oil, synthetic fuels, wood, other biomass, and wind are examples of backup sources of energy other than utility-produced electricity or gas. For many solar-heated homes which are also energy-conserving, wood combustion is a popular source of backup heat, and its impact on mitigating utility peak loads should not be discounted.

With a significant curtailment in the availability of gas, oil has become electricity's prime competitor as the backup to solar energy systems. For very energy-conserving solar-heated homes, however, electricity is often chosen as a backup because of its low initial cost, the relatively small amounts of expensive electricity that are needed, and the fact that for most cooling systems, electricity is the primary source of energy.

Systems with Gas or Electric Backup. Regardless of the backup system, a solar collector as a producer of thermal energy is analogous to an electric immersion coil in an off-peak space-heating unit: excess energy is stored as heat for use when the energy source is not available. With an off-peak electricity unit, the source is relatively predictable and controllable and can be easily programmed to provide for the entire demand of the building. For solar systems, the source (the sun) is relatively unpredictable in both its schedule and the quantity of energy it makes available. But, clearly, just as the design of the electric coil has an impact on the performance of the electric system, so, too, collector design has an impact on the performance of the solar-heating system and in turn on the interface with the utility providing electric backup.

Among the variables in the decision-making process which affect the utility interface are the size of the array; the orientation and tilt; the type of collector, whether air-heating or liquid-heating; and the operating temperature range. Among the parameters which must be considered in the decision-making process are the design integration of the various components and subsystems, the HVAC system design, the building design, the building's thermal demand profile, the utility load profile, and climate. For example, the ability of high-temperature collectors to produce high temperatures at high efficiencies at levels sufficient to power solar-actuated cooling equipment is analyzed separately from low- and medium-temperature-producing collectors.

Variation of the size of the collector is not included in the matrix, but as a solar building design option it must be given serious consideration. It has been hypothesized that in climates in which sunshine is coincident with peak demand, properly sized collectors can greatly reduce the electrical heating demand of a building on the utility company during its winter peak period. In addition, in hot climates in which sunshine is coincident with utility peaks, a properly sized collector used to actuate cooling equipment can also greatly reduce the electrical demand of the building during a utility's peak summer period. [See Feldman and Anderson (1976a). Arizona is a good case in point. This is also the case for solar hot-water heating.]

Storage of thermal energy is implicit in the interface between solar buildings and utilities. The thermal capacity of storage is the most important variable in designing for this interface. Although storage is a subsystem, it is discussed not separately as a topic of its own but rather as part of the entire system design. [For a different approach, see Asbury and Mueller (1977).]

The manner in which collectors and storage are brought into a solar system design, as well as the manner in which the solar system is integrated with electrical backup, affects the overall solar building/utility interface.

The auxiliary backup can interface with the solar system as a means of mitigating utility peak demand in two primary ways. One is by being a system unto itself, as in most conventional designs (separate auxiliary); the other is by integrating the backup with solar storage (integral auxiliary). Separate auxiliary is the conventional approach to solar building design. Such systems can mitigate peak loads on demand (conventional) and by off-peak systems. Conventional systems use auxiliary energy as required by the building and the system. They are distinguished by a distinct piece of hardware (such as a furnace). They mitigate peak loads under special conditions, such as the coincidence of sunshine with utility peak loads. Off-peak systems are those in which the auxiliary makes use of its own off-peak storage system and is integrated into an overall system in much the same ways as present on-demand systems.

Many solar water heaters are examples of integral auxiliary systems, and their design can also be applied to space heating. In these systems, an electric immersion heating coil located near the top of the liquid heat storage tank heats the liquid as needed. Alternatively, an electric furnace heats a rock solar-heat storage bin. Off-peak systems which restrict the heating of the solar storage by the auxiliary to off-peak periods are the simplest way of integrating the auxiliary with the solar, thus ensuring reduced utility peaks. The primary disadvantage of both on-demand and off-peak systems of integral storage is that they increase the temperature of the storage, resulting in potentially reduced overall efficiency of the solar equipment. Very often the heat storage system can be oversized, perhaps double or triple the size of a conventional storage unit, as a means of reducing this negative impact on performance.

An alternative method of reducing peaks is to use two heat storage containers in which the off-peak operation of the heat pump can transfer heat from one storage to the other. The low-temperature store is then circulated to the collector to be reheated while the high-temperature store is used for heating the building.

There are two primary peak-mitigating methods of using solar energy for cooling. One uses heat from the sun to actuate solar cooling equipment; the other uses the heat storage system from winter heating for storing coolness produced off-peak by conventional equipment. Again, as with heating, there are two ways of tying into auxiliary: with separate auxiliary and with integral auxiliary. However, it should be noted that with solar-powered cooling, high-temperature heat can be stored or used immediately to produce cool, the excess of which is then stored. Solar heat can also be used to regenerate dessicant material (for dehumidification) which can be readily stored.

While solar-powered cooling systems are still relatively expensive, we will see widespread applications of the use of conventional cooling equipment to cool the thermal storage that was used for solar heating during the winter, during the off-peak hours. Examples include off-peak cooling of solar storage (merely using conventional cooling equipment to cool thermal storage during off-peak hours); constant load cooling of solar storage (using conventional cooling equipment to constantly, but at a low and steady rate, cool thermal storage); nocturnal radiational cooling of solar storage (nocturnal radiators can be used during off-peak hours to cool thermal storage); and outside night air (night regenerative cooling of solar storage). In some climates, the outdoor air at night is cool enough to be stored in the thermal storage for use during the day. In most locations, however, the night air must be cooled additionally through evaporative cooling. The supercooled outdoor air is then stored in thermal storage.

Peak-Mitigating Solar Building Design Solutions:
Matrix of Their Potential Impact on Utility
Energy/Demand Balance

The matrix of the potential impact on utility energy/demand balance analyzes the various peak-mitigating solar building design solutions with regard to the utility energy/demand balance. (See table 1–1.) Although it would appear that most efforts to use solar energy and off-peak electricity should result in the conservation of energy, this is not necessarily the case. For example, a conventional off-peak electrical unit which has a capacity for storing heat for use during off-peak hours is constantly losing heat through normal heat-loss processes. This storage loss results in overall increased energy use compared with conventional on-demand electric resistance heating. On the other hand, most energy-conservation practices and most uses of solar energy will greatly reduce the amount of utility energy required by a building. At the extreme, buildings without any backup systems or with backups other than electricity will use the least amount of utility-generated energy.

Table 1–1
Peak-Mitigating Solar Building Design Solutions: Matrix of Their Potential Impact on the Utility Energy/Demand Balance

Peak-Mitigating Solar Building Design Solutions	*Factors in the Utility Energy/Demand Balance*			
	Utility Energy Conservation	*Utility Demand Reduction during Peak*	*Balance of Energy Conservation versus Demand Reduction*	*Shift of Peak Demand to Off-Peak*
Solutions Based on Architecture				
Energy conservation	+ +	+ +	SQ	SQ
Collection/Rejection				
Heating	+ +	+ −	−	SQ
Cooling				
Nocturnal radiation	+ +	SQ b	−	SQ
Natural ventilation	+ +	SQ b	−	SQ
Thermal mass	+	+	+	+

Table 1–1 continued

Combined collection and storage				
Skytherm	+ +	+ +[c]	+ +	SQ
Trombe/Baer walls	+ +	+, + +	+, + +	+
Glass in combination with large thermal mass	+ +	+, + +[a]	+, + +[a]	+
Solutions Based on Mechanical Systems				
Energy conservation	+ +	[d]	[d]	[d]
Systems without electric or gas backup				
No backup system				
Heating	+ + +	+ + +	SQ	N/A
Cooling				
Nocturnal radiation	+ + +	+ + +	SQ	N/A
Natural ventilation	+ + +	+ + +	SQ	N/A
Solar-actuated	+ + +	+ + +	SQ	N/A
Backup system other than gas or electric				
Fossil (coal, oil, synthetic fuels)	+ + +	+ + +	SQ	N/A
Wood, biomass	+ + +	+ + +	SQ	N/A
Wind	+ + +	+ + +	SQ	N/A
Systems with gas or electric backup				
Collectors				
Low- and medium-temperature	+ +	SQ[e]	− −[f]	−[f]
High-temperature	+ +	+[e]	−[f]	−[f]
Systems design				
Heating				
With separate auxiliary on demand (conventional)	+ +	SQ[f]	− −[f]	−[f]
With separate auxiliary off-peak	+	+ + +	+ +	+ +
With integral auxiliary on demand	+	SQ[f]	−	SQ
With integral auxiliary off-peak				
Heating storage off-peak	+	+ + +	+	+ +
Dual storage, off-peak heat pump	+ +	+ + +	+ +	+ +
Cooling				
Solar-powered				
With separate auxiliary on demand	+ +[g]	+	+	+
With separate auxiliary off-peak	+ +[g]	+ + +	+ +	+
With integral auxiliary on demand	+[g]	+	+	+
With integral auxiliary off-peak	+[g]	+ + +	+ +	+ +

Table 1-1 continued

Peak-Mitigating Solar Building Design Solutions	Factors in the Utility Energy/Demand Balance			
	Utility Energy Conservation	Utility Demand Reduction during Peak	Balance of Energy Conservation versus Demand Reduction	Shift of Peak Demand to Off-Peak
Non-solar-powered Off-peak cooling of solar storage	−	+ + +	+	+ +
Constant load cooling of solar storage	−	+	+	+
Nocturnal radiational cooling of solar storage	+ +	SQ[b]	−	SQ[b]
Night outside air, night regenerative cooling of solar storage	+ +	SQ[b]	−	SQ[b]

SQ	Status quo; not necessarily different from conventional design
N/A	Not applicable
−, − −, − − −	Negative impact; results in increased use of utility energy or increases utility peak
+, + +, + + +	Positive impact; results in decreased use of utility energy or decreases utility peak or shifts use from peak to off-peak

[a] The impact is climate-dependent. For example, where there's a coincidence of sunlight with the utility peaks, the impact will be positive. Where there is a noncoincidence of sunlight peaks, the impact will be negative because of the greater heat loss through the glass during those peak periods.

[b] This assumes conventional backup systems.

[c] This assumes no utility backup system.

[d] The impact here is very dependent on the design solution. For example, heat recovery systems may reduce demand by a factor corresponding to their consumption of electricity. On the other hand, heat pump systems may reduce total energy consumption, but may not reduce utility peaks.

[e] A high-temperature collector will more readily reduce utility peaks. This is particularly true for solar cooling. For example, in Phoenix, Arizona, a high-temperature collector in combination with sunlight which is coincident with utility peaks can power solar cooling equipment which will in turn reduce utility peak demand.

[f] This assumes that most solar energy systems save energy except during utility peaks. This, of course, varies with climate and with system design, and reference should be made to notes a and d.

[g] Actually with solar cooling, it should be kept in mind that except for Rankine cycle cooling, the overall coefficient of performance of the backup system for solar-actuated absorption refrigeration and dessicant regeneration is very low and requires a large amount of energy relative to conventional compression equipment. Thus the total energy savings may not be as great as might be anticipated.

Utility Energy Conservation. A minus sign or several minus signs in this column (Utility Energy Conservation) in table 1-1 indicate that the overall system will probably use more utility energy than a comparable conventional building. A plus sign or several plus signs indicate that the solution saves utility-generated energy. For example, systems which use off-peak conventional cooling equipment to cool thermal storage during the summer will use more utility energy than conventional on-demand refrigeration. On the other hand, solar-powered cooling combined with off-peak thermal storage shows a net reduction in utility-generated energy consumption and, therefore, has a plus sign.

For solar heating, most utility-generated energy is saved by a building when it uses the energy on demand; these categories have a double plus. On the other hand, the solar energy systems with off-peak backup have only a single plus because of the constant heat loss from the off-peak storage unit, resulting in relatively greater energy consumption than those systems which use electricity on demand.

Utility Demand Reduction during Peak. The main idea behind using peak-mitigating solar building design is to reduce demand during the utility's peak demand period. Virtually every entry in this category should have at least one plus sign.

The extreme case of reducing demand during utility peak loads occurs in two ways: with systems which do not use gas or electricity and with those buildings specifically designed to use electricity only during off-peak hours. The other solutions which are listed, however, will reduce demand during peaks, depending on various factors. For example, where sunlight is coincident with the peak period of a utility company, given adequate storage, even conventional solar building designs can reduce the utility peak. Some of the solutions listed, such as the architecture-based solution of nocturnal cooling, will reduce peaks in proper circumstances, such as in climates of clear night skies and relatively low cooling needs.

Balance: Energy Conservation versus Demand Reduction. This aspect of the analysis attempt to register the relative effects of a given solution on the reduction in both utility energy consumption and utility peak loads. Not all these solutions have a positive impact on the balance. A minus sign indicates that the solution reduces utility energy consumption by a greater factor relative to the reduction in peak load. Heat pumps and conventional solar building designs can greatly reduce the consumption by a building of utility-generated electricity. On the other hand, like most heat pump systems, a solar building's demand on the utility during its peak hours may not necessarily decline at all; or if it does decline, it may not decline by a factor corresponding to that of the reduction of overall electrical consumption. Such solutions, therefore, are given minus signs in the table.

A solar heating system which uses off-peak electricity, although perhaps not as energy conserving as an on-demand backup system, significantly reduces peak demand and thus is given a double plus.

Most architecture-based energy-conservation solutions reduce energy consumption and peak demand by corresponding amounts. Therefore, they are labeled SQ because they maintain the status quo with regard to the energy/demand balance. On the other hand, energy-conservation solutions based on mechanical systems vary greatly in their effect on the utility peak load, and therefore their impact on the balance also varies greatly. Buildings which use no electricity also maintain the balance, reducing energy consumption by a factor corresponding to that by which they reduce the demand.

Shift of Peak Demand to Off-Peak Periods. Although the energy/demand balance analysis emphasizes peak mitigation, it can also be desirable from the standpoint of utilities to not necessary reduce energy consumption of their generating electricity, but rather to decrease their peak loads by shifting consumption to off-peak periods. Energy conservation, for example, which reduces energy consumption and peak demand by corresponding factors has relatively no impact on this category. Energy-conservation measures may or may not shift consumption from peak to off-peak periods. However, the use of thermal mass can create this effect. Systems and buildings which do not make use of electricity at all are not applicable to this category and therefore given an "N/A."

Conventional solar heating design does not necessarily reduce the utility peak load. Such systems, however, reduce total utility electricity consumption. This reduction usually occurs during off-peak hours, and instead of shifting consumption to off-peak times, it merely reduces consumption during off-peak times. Solar building designs using off-peak electricity both reduce peak loads and, in most cases, shift consumption to off-peak periods.

Peak-Mitigating Solar Building Design Solutions:
Matrix of Decision-Making Parameters

This matrix analyzes the effects of various decision-making parameters, including site; climate; building size, type, and configuration; utility load profile; and use patterns (and comfort) on the applicability of various peak-mitigating solar building design solutions. (See table 1–2.) The analysis is based on the total system design, not merely on the particular portion of it relating to the off-peak subsystem. For example, under the parameter of site, most solar energy systems require an access to sunlight. On the other hand, this is not a requisite for the integration of an off-peak subsystem into the overall solar energy system.

Table 1-2
Peak-Mitigating Solar Building Design Solutions:　Matrix of Decision-Making Parameters

Peak-Mitigating Solar Building Design Solutions	Decision-Making Parameters				
	Site	Climate	Building Size, Type Configuration	Utility Load Profile	Use Patterns, Comfort
Solutions Based on Architecture					
Energy conservation	I	I	I	I	affected by people's behavior
Collection/Rejection					
Heating	access to sunlight	except very hot and cold	M	winter-peaking	M
Cooling					
Nocturnal radiation	L	clear skies at night	large rejection area needed	summer-peaking	L
Natural ventilation	wind helps	except humid	best for small buildings	summer-peaking	M
Thermal mass	I	M	L	for peaks of short duration (2 to 4 h)	may require small temperature fluctuations (2° to 4°)
Combined collection and storage					
Skytherm	I	clear skies at night	large area needed	I 100% solar is possible	L
Trombe/Baer walls	access to sunlight	except very hot and cold	M	for winter peaks of short duration (2 to 4 h)	may require small temperature fluctuations (2° to 4°)
Glass in combination with large thermal mass	access to sunlight	L	M	for winter peaks of short duration (2 to 4 h)	may require small temperature fluctuations (2° to 4°)

Table 1-2 continued.

Peak-Mitigating Solar Building Design Solutions	Decision-Making Parameters				
	Site	*Climate*	*Building Size, Type Configuration*	*Utility Load Profile*	*Use Patterns, Comfort*
Solutions Based on Mechanical Systems					
Energy conservation	I	I	most applicable to large buildings	I	affected by people's behavior
Systems without electric or gas backup					
No backup system					
Heating	access to sunlight	easiest in mild climates	best for small heat loads	winter-peaking	wide indoor-temperature fluctuations help
Cooling					
Nocturnal radiation	I	clear skies at night	small cooling loads	summer-peaking	wide indoor-temperature fluctuations help
Natural ventilation	wind helps	except humid	small loads	summer-peaking	
Solar-actuated	access to sunlight	clear and sunny	small loads	summer-peaking	
Backup system other than gas or electricity					
Fossil (coal, oil, synthetic fuels)	I	I	I	winter-peaking [a]	L
Wood, biomass	wood lot helps	forested	small buildings	winter-peaking [a]	M
Wind	windy	winds coincident with seasonal demand	small loads	I	L
Systems with gas or electric backup					
Collectors					
Low and medium temperature	access to sunlight	coincidence of sunlight with peaks	L	winter-peaking	L
High-temperature	access to sunlight	coincidence of sunlight with peaks	L	I	L

Systems design

Heating

With separate auxiliary on demand (conventional)	access to sunlight	coincidence of sunlight with peaks	I	winter-peaking	I
With separate auxiliary off-peak	access to sunlight		I	winter-peaking	I
With integral auxiliary on demand	access to sunlight	coincidence of sunlight with peaks	I	winter-peaking	I
With integral auxiliary off-peak	access to sunlight		I	winter-peaking	I
Heating storage off-peak	access to sunlight		I	winter-peaking	I
Dual storage, off-peak heat pump	access to sunlight		I	winter-peaking	I

Cooling

Solar-powered

With separate auxiliary on demand	access to sunlight	coincidence of sunlight with peak	L	summer-peaking	I
With separate auxiliary off-peak	access to sunlight		L	summer-peaking	I
With integral auxiliary on demand	access to sunlight	coincidence of sunlight with peak	L	summer-peaking	I
With integral auxiliary off-peak	access to sunlight		L	summer-peaking	I

Non-solar-powered

Off-peak cooling of solar storage	I		I	summer-peaking	I
Constant load cooling of solar storage	I		I	summer-peaking	I

Table 1-2 continued.

Peak-Mitigating Solar Building Design Solutions	Decision-Making Parameters				
	Site	Climate	Building Size, Type Configuration	Utility Load Profile	Use Patterns, Comfort
Nocturnal radiational cooling of solar storage	L	clear night skies	large rejection area needed	summer-peaking	I
Night outside air, night regenerative cooling of solar storage	I	dry night air	L	summer-peaking	I

I　Independent of parameter, widely applicable, can be designed to accommodate given conditions

L　Low, relatively insignificant effect of parameter on design decision

M　Medium, relatively significant consideration given to these parameters in the design decision

[a] Assumes availability of gas- and electricity actuated air conditioning only.

Site. Site has a fairly important impact on the applicability of the solar energy solution with regard to the availability of sunlight. The integration of the off-peak or other peak-mitigating systems, however, usually is not affected by the site. Solutions requiring natural ventilation may be aided by a windy site. Systems that use wood as a backup system can be helped by having a wood lot. Nocturnal radiational cooling systems are aided by a clear view of cold night sky.

Site does not have much effect on the applicability of design solutions, such as energy conservation. Site has a low and relatively insignificant effect on design solutions such as nocturnal radiational cooling since most sites provide an open view of the building to the cold night sky.

Climate. The most important impact of climate on the applicability of various solutions is the coincidence of sunlight with utility peak demand. This is particularly important when conventional solar building design is viewed as a solution to mitigating peaks. Architecture-based solutions, such as large glass areas in combination with large amounts of thermal mass in the buildings, are best in mild climates and in climates which have diurnal extremes of hot and cold. This is true particularly if these solutions are used in combination with HVAC systems to reduce peak loads. The applicability of off-peak HVAC system design to solar buildings is relatively independent of climate type as long as the solar building itself has been designed in a manner suitable to its climate.

Building Size, Type, and Configuration. The applicability of various peak-mitigating solutions is largely dependent on the building to which they are applied. For example, the creation of large glass areas in combination with thermal mass is a solution most applicable to small structures such as residences. Thermal-mass applicability decreases as building size increases. In hospitals, for example, total building demand may vary only slightly from one hour of the day to the next. Thermal mass may serve very little purpose in mitigating peak loads unless it is sufficiently large to keep the building near the comfort zone even while HVAC equipment is shut down during short peak periods.

Most energy-conservation solutions based on mechanical systems, with the exception of heat pumps, are applicable to large buildings. A heat recovery system is a good example. Buildings which require no backup energy are those with very small heating and cooling loads. These, of course, are usually very energy-conserving small buildings in mild climates. The application of solar collectors to buildings is insignificantly affected by building design. Most such systems have relatively minor impact on building design, in comparison with architecture-based systems. Virtually every size, type, and configuration of building can be adaptable to solar collectors.

If we assume that a mechanically based solar heating system has been properly integrated into a building design, then the subsequent integration of the mechanical off-peak system is relatively independent of building size, type, and configuration. The primary exception would be solar-actuated cooling systems which are probably most economically applicable to buildings with large cooling loads.

Utility Load Profile. The time of day and the time of year at which the utility experiences peaks can have an effect on the choice of a particular design solution. Most heating solutions are applicable to utilities with winter peaks, while peak-mitigating cooling solutions are most applicable to utilities with summer peaks. A problem is inherent in defining those systems which are appropriate for utilities that have alternating summer and winter peaks in consecutive years.

In addition, items such as thermal mass can affect the time of day in which a peak occurs. Thermal mass, for example, can shift the time of day in which a building experiences a peak by several hours. It can also enable a building to coast through a utility peak period without requiring additional energy input. The reason is that the heat storage capacity of the thermal mass provides any necessary heat or cooling during that period. Items such as energy conservation, wind energy, and high-temperature collectors can reduce peak loads in both summer and winter.

What is not apparent in this table is the potential adverse impact which a 100 percent solar-heated building can have on a summer-peaking utility. While the building's total energy demand is significantly reduced by the solar heating system, its peak cooling demand, with conventional equipment, occurs in spite of the solar heating system, and in many cases occurs during the utility's peak period. Thus, from the point of view of the utility, its energy/demand balance is upset because of the reduction in total energy sales without a corresponding reduction in peak load. With solar-actuated cooling technology and cost-effectiveness lagging behind solar heating, encouragement should be given to the design of solar buildings in summer-peaking service territories which (1) achieve summer peak-load reductions today by factors corresponding to a reduction in consumption of utility energy through energy-conservation practices and off-peak conventional cooling systems making use of solar thermal storage or (2) can be easily retrofitted for solar-actuated cooling in the future.

It is important to consider that many solar systems in buildings are for heating only, while most utilities are or will be summer-peaking. On the other hand, a 100 percent solar-cooled building in a winter-peaking service territory will be a rare installation, and its impact on the utility will rarely need to be considered.

Use Patterns and Comfort. The energy use patterns of the occupants of the building and the expected level of comfort by the occupants can have a significant impact on the applicability of the various peak-mitigating solutions. Even the best energy-conservation measures can be more than offset by the wasteful manner in which people operate a building. The applicability of glass as a solar collector is greatly influenced by how the users of the space perceive comfort and are willing to deal with large quantities of sunlight flooding the space. The potential for using thermal mass is best realized when occupants are willing to experience fluctuations in interior temperatures. The wider the allowable fluctuations, the more applicable is the use of thermal mass. Systems which require no backup are most applicable if wide indoor-temperature fluctuations are permitted.

Most mechanical-based systems in combination with off-peak electricity for both heating and cooling can be designed to have almost no impact on the way in which people use buildings and to accomodate almost any energy use pattern within a building.

Summary

The purpose of this chapter was to outline the basic components and principals involved in space conditioning a residence. Individually, the building, the homeowner, and the electric utility have different problems as to the requirements and benefits of various technological options. Each of these components or principals influences the eventual energy system selected.

A residence is a complex structure with many diverse requirements and uses. Many opportunities, including a range of architectural, engineering, and attitudinal techniques, exist within new building technology to reduce space-conditioning needs and/or electric bills. The complexities and interaction among all aspects of a residence make it inappropriate to separate the needs of one function, such as space conditioning, from other aspects of design and utilization.

The building owner eventually is responsible for the selection of the type of design, HVAC system, and operation of the home. The individual consumer attempts to minimize costs within his own definition of cost-effectiveness and comfort level. Because of the complexities of options and intervening barriers, the consumer is not always able to achieve complete cost minimization. The homeowner is still cognizant of tradeoffs that may exist between the various technological options.

The public utility that serves as a competing and/or auxiliary energy source is not an unconcerned party in building design and energy use.

Building loads, in both annual kilowatt hours consumed and kilowatts demanded, determine utility operation and costs. The rates in part influence building configuration and operation.

The qualitative assessment of the solar-utility interface illustrates the complexity of the issue. The impact of solar technologies on electric utilities cannot be generalized but must account for different types of applications, sizing, operation, and building design. In addition, the siting, climate, and utility load profile all determine the ultimate impact.

Notes

1. One quad $= 10^{15}$ Btu. For the contribution of residential space conditioning within total U.S. energy consumption, see Federal Energy Administration (1974, p. 32), and Energy Fact Book (1976).

2. The effective temperature of the exterior should not be represented by ambient temperature alone. Sol-air temperatures were developed to account for radiation; see B. Anderson (1977, pp. 306–307). The sol-air temperature is an approximation of the combined effects of solar radiation and convection, producing a measure of the experienced surface temperature.

3. See Trombe et al. (1976, 201–208) and B. Anderson (1977, pp. 91–101).

4. See Baer (1975) and B. Anderson (1977, pp. 102–105).

5. See Hay (1971) and B. Anderson (1977, pp. 111–114).

6. See Berlin, Cicchetti, and Gillen (1974, pp. 1–11) and Olson (1975) for a general discussion of electricity history and present problems; see also Turvey (1968).

7. This phenomenon of overinvestment by public utilities is commonly referred to as the "A-J-W effect" after authors Averch and Johnson (1962) and Wellisz (1963); see also Kahn (1970) or Priest (1969). See the discussion in last chapters of this book.

8. Power blackouts refer to sudden drops in supply due usually to equipment failure; service curtailment refers to long-run supply shortages. Tolley, Upton, and Hastings (1977) identify costs of electricity curtailment as either further quantity restrictions or rate increases and specify costs to the consumer as well as regional unemployment costs. The cost of power blackouts can range from minor inconveniences to major disruptions of large urban areas (EPRI 1978b).

9. Berlin, Cicchetti, and Gillen (1974) credit not only fuel costs and plant size but also plant age and location as factors affecting variations in incremental cost per kilowatt/hour. For example, in a large eastern utility the incremental fuel cost ranges from 3.3 mils/kWh to 29.51 mils/kWh.

10. A summary of state of the art of electric utility storage can be found in Feldman and Anderson (1976b, pp. 108–130) and Huse et al. (1976).

11. Experiments, although controversial in design, resulted in shifts in electricity usage from peak to off-peak periods. There is less evidence of hardware purchase probably in part because of the experimental nature of such rate schedules; see Northeast Utilities (1977). For an excellent review of load management issues see Morgan and Talukdar (1979).

2 The Systems Model—A Conceptual Overview

Many economic studies of the cost-effectiveness of active solar energy hot-water and space heating systems have been performed. These reports have provided evidence of the economic viability of solar technologies, described consumer design optimization of solar systems, indicated potential market penetration of solar applications, and suggested proper government policy for solar development and energy independence.[1]

The first rigorous study on the economics of solar hot-water and space heating systems was performed by Tybout and Löf (1970).[2] Their use of computer simulation to predict solar performance and cost-effectiveness was a novel approach in the design of solar systems, generating findings to show the potential development of solar heating in most cities in the United States. Using some general approximations of collector performance, collector losses, and storage losses, the model calculated the amount of useful energy obtained from available solar radiation. This program was then used in conjunction with cost estimates to arrive at optimized parameters for collectors and storage units. Whether appropriate or not, many of the findings from this study and several updates today represent the most commonly used rule-of-thumb estimates for proper sizing of the storage and collector units and the proper tilt of the collector.

One significant criticism of the work of Tybout and Löf involved their estimation of collector costs. They assumed that mass production would lower collector costs to $2 to $3 per square foot. A subsequent report by the University of New Mexico (Ben-David et al. 1977) reinvestigated the cost-effectiveness of solar heating and hot water, utilizing higher estimates ($3,505 fixed cost and $9.66 per square foot for variable costs) of solar costs and conventional fuel prices.[3] Realizing the inevitability of higher future prices of fuels and electricity, and the subsequent importance of accounting for these price rises in solar economic studies, this study incorporated a life-cycle cost analysis which credits savings over the entire operation of the solar heating system. Using financial analyses along with a set of precalculated performance parameters to predict solar feasibility, they also found that given certain tax policies, price decontrols, and interest rates, solar energy "could become widespread by 1990."

The critical importance of the financial parameters, particularly interest rates, fuel costs, and solar system costs, on the final outcome of these feasibility studies has led to several additional reports. However, the uncertainties in future costs have created different predictions of solar economic feasibilities, as is evidenced by an optimistic report toward solar cost-effectiveness by Bennington et al. (1976) and a report with a similar methodology which was more pessimistic.

The uncertainty of these financial parameters has led to confusion over the true position of solar water and space heating as a future energy source. However, the emphasis on reworking these cost estimates has diverted attention from several more serious methodological problems, lessening the value of the conclusions of the reports.

Problems with Previous Solar Economics Studies

Analyses of the consumer cost-effectiveness of solar water and space heating systems compare the solar collection system costs against conventional fuel bills. The fact that the expenses of solar space heating and domestic hot-water systems compete favorably with present electricity bills does not ensure the appropriateness of solar as an alternative to electricity. Optimization must involve all the heating and cooling aspects of the residence and the interaction with utility load factors and costs. Energy conservation and passive solar designs must be evaluated as alternatives.[4]

Two reports issued by the National Bureau of Standards described several important economic criteria required to achieve true optimal building configuration. Peterson (1974) pointed out the need for marginal analysis. Marginal analysis is concerned with the costs and benefits that are derived from the marginal, or next, unit of investment rather than the average benefits and costs. Optimal investment criteria require the equating of the ratio of marginal benefits to marginal costs for active systems with the ratio of marginal benefits to marginal costs for every other investment. A second influential report by Ruegg (1976b) qualitatively illustrated the sensitivity of economic parameters such as discount rate and fuel costs to the cost-effectiveness of particular investments. More recent work by Duffie, Beckman, and Brandemuehl (1976) and Brandemuehl and Beckman (1978) has quantified this relationship.

Recently several studies have reported on the cost-effectiveness of passive solar energy systems. Many of these reports focused on an individual experimental system and are specific to the details of that system. Other reports—Kohler and Putnam (1978), Taff et al. (1978), Balcomb, Hedstrom, and McFarland (1977), and Balcomb and McFarland (1978)—have relied on simplified algorithms for calculating a generic passive system. Roach et al. (1978) utilized the simplified passive technique of Balcomb to draw comparisons between these passive findings and the earlier New

Mexico reports on active feasibility.[5] They found that the passive designs examined (Trombe walls) can be more cost-effective than active air collector–rock bed storage systems.

The Roach study utilized a standard wood frame and design with a fixed level of insulation, approximately equivalent to the American Society of Heating, Refrigerating, and Air Conditioning Engineers (ASHRAE) standard 90–75.[6] In no study has the specification of insulation been optimized in conjunction with solar investment. The modeling of the physical heat flow in the previously mentioned reports is grossly unsophisticated. Most of the studies did not utilize simulations of heat transfer but used generalized design charts which provide average monthly performance.

In all the studies mentioned except for that of Roach et al., a solar performance model such as "F-Chart" was used to evaluate solar performance. The f-chart is a statistical compilation based on more detailed solar simulations developed by the University of Wisconsin Solar Laboratory (Beckman, Klein, and Duffie 1977). Simulations of many solar collector/storage combinations were performed using TRNSYS, a transient simulation that models heat flow through each system component to arrive at an hour-by-hour measure of collector, storage, and overall system performance. The results of these simulations were compiled by using regression analysis to derive an equation which determines solar performance as a function of collector size and storage size. In this way it is possible to specify average monthly climatic values for temperature (degree-days) and insolation as an indication of solar system monthly performance.

The results of the f-chart analysis are constrained by the limitations of the original regression analysis. The validity of f-chart does not extend to variations in the specification or operation of the collector system. Also, f-chart accuracy is limited in its ability to predict other aspects of system performance such as the performance of solar heat gain through windows and the effects of incremental insulation on thermal mass. The dependence of all these simple models on monthly values of weather and loads makes determination of load variation by time of day impossible.

The variation in temporal demand for heat energy and auxiliary energy is a particularly important factor with regard to the source of auxiliary supply. Variations in load for electric utilities account for large variations in cost of service. These cost differences have necessitated experimentation with tariff schedules where price varies with variations in load. Cost optimization for the individual consumer must account for these potential changes in utility tariffs.

The entire solar-utility interface is ignored or misconceived in previous solar economic reports, for the impact of alternative tariff designs on solar economic feasibility is not considered. In addition, the impact of solar water and space heating on the servicing auxiliary source is not incorporated into economic studies of solar feasibility, despite several reports that indicated serious impacts and barriers (Feldman and Anderson 1975a, 1976a, 1976b; Lorsch 1977). Ignorance of costs of service which for electricity are not directly incurred by the consumer at present can cause problems for consumers when the utility redirects these costs to them. Anticipated electricity savings calculated with present tariffs may not actually be realized under these new tariff designs.

Perhaps the greatest weakness of these economic studies is the identification of consumer cost-effectiveness as a measure of national economic efficiency. Consumer optimization does not necessarily lead to societal optimization when inefficiency in the form of externalities exists in the marketplace. The external costs incurred and tax subsidies received by the utility and not covered by the solar consumer's electric bill are an example of this market imperfection. Several other costs are external to the individual's cost-effectiveness decision but affect significantly the value of solar energy to the nation as a whole. Among these costs are the social cost of pollution from conventional energy sources, the cost of energy dependence, and the impact on foreign trade deficits.

Recent public policy for solar energy has instituted tax credits for solar energy development to partially offset any subsidies extant in other fuel sources and to promote the development of solar energy. The evaluation of the benefits of these tax credits has been directed toward the effect on consumers, a subsequent development of preferred solar technologies. What is needed, however, is an evaluation of the effectiveness of these policies in promoting the more important goals of national economic efficiency and curtailment of dependence on foreign energy. Tradeoffs that exist for the individual consumer in the use of active solar systems, passive solar systems, and energy conservation will exist for the nation as well. Market imperfections in the form of externalities cause the optimization process to differ between these two perspectives. One role of government policy is to correct these imperfections so that individual decisions are in line with national well-being.

Thus far we have described the basic components/principals involved in the space conditioning and hot-water heating of a residence. Individually the building, the homeowner, and the electric utility possess different problems as to the requirements and benefits of various technological options. Separately, each of these component/actors influences the eventual energy system selected.

A residence is a complex structure with many diverse requirements and uses. Many opportunities including a range of architectural, engineering, and behavior-modifying techniques exist within new building state-of-the-art technology to reduce space-conditioning needs and/or electric bills. The complexities of and the interaction among all aspects of a residence make it inappropriate to separate the needs of one function, such as space conditioning, from other aspects of building design and utilization.

The building owner eventually is responsible for the selection of the type of design, HVAC system, and operation of the home. The individual consumer attempts to minimize costs within his own definition of cost-effectiveness and comfort level. Because of complexities of options and intervening barriers, the consumer is not always able to achieve complete cost minimization. The homeowner is still cognizant of tradeoffs that may exist between the various technological options.

The public utility that serves as a competing and/or auxiliary energy source is not an unconcerned party in building design and energy use. Building loads in both kilowatt hours consumed annually and the distribution of kilowatts demanded determine utility operation and costs. The rates in part influence building configuration and operation.

The rates presently utilized by the electric utilities do not represent the true social cost of service for electricity. Furthermore, other costs remain external to either the homeowner or the utility. Among these costs are the impact on balance of trade, impact on energy independence, impact on the environment, and various subsidies to the industry. Costs to the nation such as the balance-of-trade costs are the direct result of the consumption of fossil fuels.

A study of Battelle Pacific Northwest Laboratories has demonstrated the extent to which government subsidies are present in competing forms of energy.[7] The lack of "fully assigned" costs may lead to an inefficient allocation of our scarce resources. Government policy could either correct inconsistencies in the pricing mechanism or continue to absorb these external costs.

It is our intent to pursue a set of objectives that are congruent with defining the quantitative nature of the utility-solar interface. First, we construct a model of the solar-utility interface which overcomes some of the pitfalls of existing models. Initially, we examine the economic viability of space conditioning and hot-water heating by solar energy systems. However, as will be demonstrated, subsections of the model can be used for the assessment of most decentralized technologies. The model attempts to account for energy conservation and passive solar design as well. Second, we identify the impact of solar energy and energy-conservation investment on electric utility loads and finances. Third, we suggest the impact of taxation policies on the cost-effectiveness of solar systems, and the consequent effect on the solar-utility interface. After these tasks have been accomplished, one can examine the effect of various rate schedules for backup energy on solar and nonsolar customers and the rates of return for society of various investment alternatives in decentralized and centralized technologies.

Description of the Model

In order to satisfy our objectives, we use a model which examines the cost-effectiveness of alternative building designs and decentralized energy systems and measures impacts on the electric utility. The model first evaluates the performance of a building and its energy appliances; that is, the demand for electricity is calculated under average and extreme weather conditions (peak period) for one location in the utility service area. The impact of these demands on the electric utility is then measured by calculating the increment in long- and short-run costs incurred as a result of the addition of one building in the service area. Economic efficiency comparisons are then made between buildings of different designs from the points of view of the utility, the building owner, and national economic efficiency.

Case Study Sector

Utility Selection. Feldman and Anderson (1976a) have concluded that the solar-utility interface impacts are rather site-specific. Analyses must be performed on an individual utility basis, since variations in the ambient weather conditions, load curves, and generation mixes of utilities will be the prime determinants of the magnitude of the impact. Therefore, one must select the electric utility to be examined on the basis of the jurisdiction in question. Utility results cannot be transferred to other utilities for policy purposes; each utility must be analyzed individually.

Building Performance Sector

Building Selection. The development of a well-defined building model is one of the prerequisites of this methodology. Special care has been taken to account for many factors which affect building loads.

The National Association of Home Builders Research Foundation, Inc., has specified building parameters typical of new building construction (Westinghouse 1973). Conformity to these values has been attempted as much as possible with only slight modifications to aid in ease of modeling and adaptation to solar equipment. The building modeled is a 1500-ft² single-story ranch 18 m [58 ft] long, 8 m [26 ft] wide, with 2.4-m- [8-ft-] high walls. The building is oriented with the major dimension south. The building has a basement and a detached garage. Beyond these specifications, many other factors which affect the energy consumption of a typical residence are specified: insulation, wall structure, active solar, passive solar, thermal mass, fenestration, control devices, and human behavior.

Building Performance Simulation. The performance of each building within the utility service area is modeled using TRNSYS, a transient computer simulation model developed at the University of Wisconsin Solar Energy Laboratory at Madison. Several modifications to TRNSYS, including ones that enable the modeling of direct-gain passive solar, were developed.

Specify Typical Weather Year. The thermal performance of a residential building is largely a function of weather. In order to simulate performance by using TRNSYS, hourly weather data are needed. Annual weather data consisting of hour dry bulb, wet bulb, radiation, and wind speed are compiled by the National Climatic Center for a number of locations. This data base is constantly being expanded.

Average Annual Electrical Consumption. The results of the run of TRNSYS driven by the annual weather sequence yield the average yearly performance of each building. Performance of a building is measured by the amount of electricy demanded. Because this electricity is supplied by the electric utility, it is important to gauge the performance of a building by its effect on the utility's load curves. These performance calculations are the average annual consumption of electricity broken down by each building. It should not be thought of, however, as a complete account of the building's performance or the complete effect on the utility, since the values represent space conditioning and hot-water heating only. The model assumes that the remainder of a household's electric load, for lighting, for example, is the same in all buildings and can therefore be ignored in the analysis.

Peak Weather Sequences. Average electrical consumption provides an incomplete analysis of building performance because extreme weather conditions are not emphasized. This is important, since extreme weather conditions are also responsible for the utility's peak loads. The peak loads determine the capacity requirement and subsequent capital costs to the electric utility. These costs to the utility may be more accurately accounted for in new tariff designs. For this reason, it is necessary to include the performance of the building at the time of the utility system peak as part of the overall performance of each building.

In order to test the performance of a building under peak-load conditions, five-day weather sequences were compiled. These weather sequences coincide with the historic times that the utility experienced its peak load. The last day of these sequences is, in fact, the peak-load day for a given year. It is assumed that five days is sufficient to mitigate the influence of initial starting conditions (say, storage tank temperature) in the simulation of solar building performance. The critical factor is the performance of each building during the peak period on the peak-load day. During this period incremental electricity demand causes the need for additional capacity for the utility.

Local climatological data (LCD) sheets from the National Climatic Center and weather tapes provide data for the service area. Five-day sequences from these data corresponding to the utility peak period are used in the building performance simulation.

Peak Electrical Consumption. The performance of each building at the times when the utility experiences peak demands can be compared, and the kilowatt demand of each building can be determined.

Utility Finance Sector

Short-Run Cost. The performance of each building has been determined as above. The impact of each building on the electric utility can be analyzed by examining expenses incurred by the utility and broken down into three components: energy costs, capacity costs, and customer costs. Marginal-cost programs determine the utility's long- and short-run incremental costs by examining separately the contribution of each component. The model examines the effect of an additional building on the load of the utility. This direct analysis permits one to ignore any common expenses between different building configurations other than for heating and cooling. In this manner the direct contribution of space conditioning to the expenses of the electric utility is calculated.

A primary component of an electric utility's costs is its energy costs or short-run costs. Short-run costs include all costs that vary with the amount of energy produced—the cost of fuel and operation and maintenance costs that vary with output. The short-run costs per kilowatt hour increase as output increases because higher output normally requires use of less efficient generating plants. This means a greater use of fuel and higher cost per kilowatt hour. Should output drop to zero, however, there would be no short-run costs.

The demand in kilowatts of an electric utility is met at any time by a group of generating plants whose total capacity is sufficient to meet demands. Each of the generation units making up this capacity has a fuel requirement per kilowatt hour, which is determined by the plant's thermal efficiency. The cost of that fuel—gas, oil, coal, or nuclear—is the incremental fuel cost of that plant. The utility operates a system of economic dispatch where available plants with the lowest incremental fuel costs provide the necessary capacity, so as to minimize total costs.

While there exist sophisticated computer programs to derive the short-run marginal costs for each hour of the year, it is possible to generalize plant efficiencies into broad categories so as to provide a good representation of short-run marginal costs. In this book, incremental fuel costs are derived from the weighted average of historic fuel costs over time-of-use demand periods as calculated by the utility. The demand period segregates groups of hours which normally have peak loads from those which experience lower levels of load. In addition, a shoulder period or partial peak may also be specified. The incremental fuel cost for each period is adjusted by the appropriate, calculated loss multiplier for transmission energy, to account for losses between the point of generation and the point of use. Variable operation and maintenance costs per kilowatt hour are added to calculate the short-run marginal costs, which are multiplied by the number of annual kilowatt hours in each of the demand periods to derive the annual short-run cost to the utility. This process is repeated for each building.

The calculation of average short-run costs is considerably complicated when a significant portion of a utility's capacity is derived from hydropower. Supplied hydropower has short-run marginal costs near zero, but since it is far less reliable (not always available) compared to other types of plants, annual fuel cost is closely related to availability of hydropower. The incremental fuel cost is, therefore, normally based on some average hydropower availability.

Long-Run Costs. The second aspect of a utility's costs and the effect of solar buildings on those costs is the demand or capacity cost. This cost is a result of the demand for additional generating capacity and subsequent transmission and distribution needed to service the new growth. These are opportunity costs which arise from an unexpected increase in demand (Berlin, Cicchetti, and Gillen 1974)—or an unexpected decrease in supply. If the marginal-cost theory is used, the cost of meeting this demand should be paid by those users who instigated the need for additional facilities, that is, as a function of the load placed on the system by each at the time that the system has the least reserve capacity. The long-run cost of the utility is the annualized capital cost of the capacity required by each of the buildings. The costs are based on utility-supplied data on marginal capacity costs.

The model calculates the effect of one additional building on the utility peak loads. What is determined is the increment to utility load of the building at the time the utility has experienced its peak demands in the past. The cost of this increment to the existing load from each building can be calculated by using data from the utility on its incremental generation costs, supplied by the utility or derived, for example, from Cicchetti, Gillen, and Smolensky (1976). The effect on the capacity requirements of the utility of each building can therefore be compared, where the cost in dollars for new capacity required is the common denominator.

By using TRNSYS, each building is simulated for five days prior to each of the past five annual peak-load days. The performance of the building on the peak day of the utility is obtained. The desired result of the simulation is the diversified electrical demand for the peak period (the hours for which there is probability of loss of load) on the peak-load day. The demand is considered diversified over time, since electric draw for space conditioning and water heating occurs in short bursts. When this demand, however, is integrated over a one-hour period, the load profile is smooth and reflects the heating or cooling load of the buildings. The model uses the average demand of the building during the peak period of the peak-load day for each of five years as a surrogate for capacity requirement. It is not useful to specify the demand of the building during the instant of system peak because of the lack of precision in the building model and in utility forecasting.

The long-run cost to the utility for each building is derived by calculating the marginal annual generation capacity cost and the marginal annual transmission-and-distribution capacity cost for the utility and multiplying the total annual capacity cost by the kilowatt requirement of each building at time of system peak.

Rate Schedule of the Utility. Electric utilities often have a number of rate schedules, which apply to different classes of customers. As a result, the selection of the proper rate for the heating and cooling portion of the electric bill is a complicated process. Under average cost pricing, most utilities utilize a declining block rate. As consumption increases, the unit price per kilowatt hour decreases. Because heating and cooling consumption may be viewed as an increment to other uses of electricity and since we want to isolate the marginal effects of various configurations, the appropriate block to bill for the heating and cooling kilowatt hour demand is at the marginal rate for average monthly use. We assume that a typical Colorado household uses approximately 500 kilowatt hours per month for lighting and so on, exclusive of electricity for space conditioning and hot-water heating.

We assume that an appropriate rate for billing incremental kilowatt hours is the price of the 501st kilowatthours (P_{501}) for the existing declining block rates; the customer's bill is computed from $-(KWh \times P_{501})$. A time-of-use rate was calculated for Colorado Springs in the absence of any existing or proposed tariff. The rate schedule is developed from Cicchetti, Gillen, and Smolensky (1976) and represents a potential short-term implementable tariff.

Electric Bill. The calculation of the electric bill is simply a matter of multiplying the appropriate rate times the amount of consumption. In the case of time-of-use rates, the hour-by-hour consumption of electricity for each building is aggregated into the specific demand periods.

Revenue to Utility. The revenue to the utility is specified as the amount equivalent to the electric bill of the building owner. This represents only the revenue from heating and cooling uses of electricity.

Total Incremental Costs to Utility. The total costs are merely the sum of the short- and long-run costs. As in the case of revenues, this value is meant not as the total cost but as a mode of comparison between buildings. Since other costs and revenues are ignored, it is not possible to derive to what extent total revenues and total costs are out of balance.

Building Owner Sector

Incremental Construction Cost. The costs of all building options have been calculated using *Means Construction Cost Data* (Means 1976). The relevant value is the additional cost of each building above a base value—a 2 × 4 in-stud-wall, fully insulated standard building. Comparisons of the cost-effectiveness of each building can be made under life-cycle or first-year cost analysis. The effects of tax credits can also be evaluated.

Economic Efficiency Sector

Economic Efficiency. The determination of the economic efficiency of solar energy construction must take into consideration the effects on three different factors: the solar building owner, the utility, and national economic efficiency. For the solar building owner, the important economic consideration is whether the annualized capital cost of the solar building is exceeded by the average annual savings in the heating-cooling bill within a specified payback period. For the utility, the total incurred to supply energy to any customer should be recouped by the revenues from sales to that customer. If this is not the case, then either other utility customers must subsidize that customer or the utility must devise a rate which puts costs and revenues in balance. In practice, the pursuit of equity is such a difficult process that rates are derived to balance costs and revenues over broad classes of users. National economic efficiency will be served if the total costs to society of the solar construction are less than those of substituting some other form of energy for solar energy. These total costs must include the incremental costs for the solar energy systems and the potential savings from delayed utility capacity and accompanying fuel savings.

The strengths of the model described here lie in several advantages over the models presently used. (For a review of solar-utility interface models see Flaim et al. 1979.) For example, congruent with microeconomic theory, long- and short-run marginal costs are accounted for in examining the impact of solar on utilities. Only the work by Bright and Davitian (1978) on solar hot-water heating in the Long Island Lighting Co. uses a long-run marginal-cost analysis. They employ WASP II, a dynamic optimization model that calculates annual system variable costs for a number of utility system configurations subject to certain reliability constraints. The generation expansion plan which minimizes the present value of total system costs over a given planning future is selected. The marginal cost of supply electricity for backup power is the difference in these total costs (with and without specified solar market penetration) divided by the difference in total electricity generated. In the model, transmission and distribution costs are not considered by WASP II. WASP II is an exceptionally cumbersome model which, to our knowledge, is not used by any U.S. utilities for actual planning purposes. This is a major drawback to its use as a policy tool.

The solar performance model most widely used by investigators of the utility interface is TRNSYS. Bright and Davitian (1978), Lorsch (1977), and Debs (1979) all employ this model. None of these studies include analysis of the tradeoffs of energy conservation or passive solar design with active SHACOB systems and utility load management.

Another performance model, not yet the subject of the rigorous verification of results that TRNSYS has encountered, was developed by A.D. Little, Inc., for the Electric Power Research Institute (EPRI) and is called the EPRI Methodology for Preferred Solar Systems (EMPSS). This model fails to assess the implication of conservation measures and passive applications.

More recently the Aerospace Corporation (1978) study examined four advanced alternative systems: load management, direct solar, direct solar with load management, and solar-assisted heat pumps. By employing market-penetration assumptions and synthetic utility models, impacts on seven utilities were determined and preferred advanced alternative systems identified for displacing conventional systems of resistive heating, heat pumps, and load management. A major deficiency in the methodology is the parametric approach taken in defining SHACOB system designs, which leaves no opportunity for optimizing technical configurations or solar design fractions. Solar design fractions of 0.25, 0.50, and 0.75 only are examined for the four fixed system types. By evaluating system performance and economic impact under this constraint, it is concluded that in the preferred systems the solar collectors are typically small—providing 0.25 solar fraction. The conclusion requires qualification. A solar fraction of 0.25 is not necessarily the optimal design. It is, however, the preferred of the three fractions examined. It should be noted that other research using an optimizing approach has found preferred systems with a solar design fraction of 0.5 to 0.7.

Another difficulty with the EPRI study was the use of synthetic utility models. If we consider the objective of identifying preferred systems for specific utilities, it would seem more appropriate to use specific utility data rather than generalizing the utility by using an EPRI synthetic utility model. Clearly some degree of accuracy was lost.

The failure of these studies has been in not optimizing and examining alternative technical configurations with regard to the solar appliances, the building characteristics, and the potential for integrated SHACOB/load management. In addition, passive solar buildings and other energy-conserving technologies, such as heat pumps, did not until very recently receive the attention they deserved with regard to mitigating negative impacts at the solar energy/utility interface.

Modeling Buildings and Solar Energy Systems [a]

In order to determine the relative importance of each building component toward saving energy, it would be necessary to build and/or monitor a vast number of buildings. The use of simulation allows for the determination of building performance at much lower investments of research time and money. The state of the art of modeling buildings is enhanced by the use of computer programs which allow speedy resolution of complex calculations.

Of primary importance to the rapid adoption of SHACOB is the adequate demonstration of the cost-effectiveness of such systems. There are thousands of operational solar-heated and several hundred solar-cooled buildings nationwide, but the performance of these buildings has not yielded conclusive results as to their cost-effectiveness. Many of the buildings are experimental either in construction or in SHACOB system design. What is needed before SHACOB becomes an important element in energy supply is a reliable body of evidence which can be used as a tool to test the cost-effectiveness of different designs.

The methods which may be used to evaluate the performance of SHACOB systems include the monitoring in real time of a solar building,

---

[a] This section was written by Richard Howard, Eliot Wessler, and Robert Wirtshafter.

monitoring in real time of an analogue building, and simulating the performance of a building by using some sort of mathematical techniques. In this section we discuss the use of simulation models in determining the performance and cost-effectiveness of SHACOB systems.

Description of the Models

The increased interest in solar energy technology recently has created a minor explosion in the availability and use of simulation models specifically designed for solar energy applications. Graven (1974) concluded that "No generally accepted programs are available to the public which can provide the desired detailed design information." Yet today simulation for solar building runs the gamut from extremely complex research to simple design programs. The appropriateness of a particular model is determined by not only specific modeling limitation, but also existing tradeoffs between facility and accuracy. The most detailed surveys to date are presented in Science Applications Inc. (1978) and Arthur D. Little, Inc. (1979).

Simple Method for Calculating the Building Load. In the simplest models, the heating (cooling) load is often merely the equivalent to the heat loss (gain). The amount of energy that leaves the building is crudely calculated via a UA value for the home, where the UA for each surface is the product of the coefficient of transmission of heat U and the area of the surface A. The UA of a building is the summation of all building surface UA's. A simple approximation of the heat loss and thus the load is found by multiplying the building UA by the total ΔT for the period.[8]

Problems with the Simple Models. While this method of heat-loss calculation is adequate for approximating average monthly or yearly energy flows, it is in many ways inadequate. The lack of complexity in the heat-loss calculations makes it difficult to differentiate the contribution of one specific building parameter from another. A large deficiency in this methodology is the fact that the methodology oversimplifies the building's operation to the extent required for proper public-policy prescription. Quite obviously, the average monthly calculation provides no indication of temporal or peak demands. Calculation of the energy bill under a time-of-use or Hopkinson tariff is impossible.

There are, in fact, other significant fluctuations in energy flows which drastically affect loads and which negate the value of average performance. The human occupant contributes to energy inputs merely by living within the home. His occupation of the home is not constant; neither are his use of hot water and other appliances nor his maintenance of internal temperatures constant.

The average weather statistic *degree-days* provides an average temperature demand calculated by subtracting a base value minus ambient temperature (usually $65°F - T_a$) for the residence. Unfortunately, the sizing and operation of the HVAC system must be capable of handling the worst case, or "design" conditions. If parts of the overall building system, for instance the solar energy system, are designed for average conditions, the entire system must still survive the extreme events. Those systems designed for only average situations exert further pressure on the entire system during the design conditions. Optimization of a single portion of the total load does not necessarily optimize performance of the system.

Another assumption of the simple *UA* model is that the load in any one period is determined by the amount of heat loss in that period. In actuality, the building distribution of heat losses and gains into the home is not an immediate process and is associated with time lags.

More Sophisticated Modeling of Building Loads—Transfer Function. The internal temperature of a building is based on a complex combination of heat flows. Convection heat flows are transmitted into the room by air circulation. There is conduction through the walls, attic, and floor, and there is radiation from other bodies. Heat is also added by internal sources including lights, appliances, and people. Accurately modeling this situation is further compounded by the constant variation in solar radiation, wind, human behavior, and temperature, all of which alter heat flows. To accurately determine the building load, it is necessary to solve simultaneously each of these heat flows through each surface of the building.

The National Bureau of Standards has developed a building load program which performs this function (Kusuda 1976). The computer algorithm solves simultaneously the heat response through each surface and from each source, and thus it is called a *response factor method*. Unfortunately, the complexity of this solution is costly in time and money despite the use of high-speed computing.

The American Society of Heating, Refrigeration, and Air Conditioning Engineers Task Force on Energy Requirements has recommended a simpler method for calculating building loads (ASHRAE 1975b). Mitalas and Stephenson (1967) developed a method utilizing transfer function coefficients to greatly simplify the computation. The transfer coefficients are developed by running the building model for a short time under the more rigorous load response factor method explained above. The results of this test are compiled into a set of weighted values which represent the generalized energy response of the walls, the roof, the attic space, and the room. The calculation of energy transfer utilizes these weighted values to calculate the distributive-lag effect of heat flow.

The emergence of many methodologies for solar energy and building performance confuses model selection. Generally it is convenient to divide these simulations into groups, depending on their complexity and usefulness. Perhaps the most widespread use of load models is the simplified design tools. These programs provide estimates of annual and/or monthly building load, solar energy performance, and energy use. The f-chart is employed most often (Beckman, Klein, and Duffie 1977). Others including Mass Design (1976) and Solcost (Hull and Giellis 1978) have not had sufficient validation to merit widespread adoption. A second group of programs including TRNSYS was developed principally to model solar energy but does contain some building modeling capabilities. Many of these programs incorporate similar algorithms for calculations. A report by Maybaum (1978) shows reasonably similar results (within 5 percent) were obtained from each model when a specific solar energy system was simulated by TRNSYS, SYMSHAC, SOLSYS, CAL-ERDA, HISPER, and LASL.[9] One of the unknown questions regarding these principally solar simulation models is the level of accuracy in building load prediction.

The use of thermal transfer functions by TRNSYS does improve its building model capabilities. There do exist some building models—BLAST (Hittle 1977) and NBSLD—which model multimode building loads more completely. Whether this level of accuracy is required of single-family residential structures is yet unclear.

Modeling Passive Buildings. One of our hypotheses is that solar energy collection may be more cost-effective when utilized in a form other than active collector system. To this point quantification of this hypothesis has been hampered by the fact that passive systems, which generally incorporate design of building, are difficult to model. The difficulties incurred with the transfer coefficients are exacerbated by energy flows normally associated with passive designs. Natural convection systems of thermosiphoning, Trombe walls, roof ponds, and direct solar gain are unknown quantities in regard to accurate performance models. The initial attempt at monitoring these buildings is being conducted by Los Alamos Scientific Laboratory.[10] At this time, however, construction practices still rely on rules of thumb developed from personal experience. Even when rule-of-thumb procedures have some basis in fact, these construction practices also rely on accounting for average performance. Analyzing these buildings during peak conditions has not been attempted.

Without adequate performance knowledge to verify modeled results, it is impossible to quantify the appropriateness of particular passive techniques or sizing configurations. The modeling must rely on algorithms developed for other similar configurations. Modeling of the more complex passive systems (complex in terms of heat flow, not construction difficulties) does not possess a known level of confidence.

One method of passive design—direct gain—can be modeled with only slight modifications to TRNSYS.[11] Direct-gain energy flows refer to the use of fenestration and building construction to collect solar radiation. The sizing and number of glazings of the windows, shading, and use of thermal shutters determine the amount of solar radiation that enters the living space. The construction weight and transfer function procedures dictate how that energy is incorporated into the load. It is unclear how well suited TRNSYS is to model these flows. The use of a more accurate building model may increase modeling confidence.

By far the most widely known and used transient system simulation program for solar energy systems is TRNSYS. Klein (1976) reports that TRNSYS has been distributed to approximately 150 companies, universities, and government agencies both in the United States and abroad. It is also the simulation model which is often cited in the literature and for which there has been the greatest amount of verification of results. It was developed by the University of Wisconsin Solar Energy Laboratory at Madison under dual funding by NSF/RANN and U.S.D.O.E. A useful reference to this program is the TRNSYS manual by Klein et al. (1973).

Also TRNSYS may be used to evaluate the thermal performance of a nearly infinite number of configurations of SHACOB systems. This capability is due to the structure of the program, which allows the specification and connection of component models to solve a set of simultaneous algebraic and differential equations describing the system. While TRNSYS may be useful in the specification of any thermal transient system, it has special application to solar energy systems because of the availability of a library of component models which most often are associated with solar energy systems.

These components include flat-plate solar collector, stratified fluid storage tank, heat exchanger, on/off auxiliary heater, space-cooling load and air conditioning, rock-bed thermal storage, liquid collector/storage subsystem, air collector/storage subsystem, domestic water-heating subsystems, and, in the newest version, passive systems. Each of the component models in the library is formulated as a separate FORTRAN subroutine. New component models are periodically added by the authors and are included in the library. In addition, the user may specify his own component model (say, for novel equipment) which can be easily added to the program.

TRNSYS is a complex and sophisticated program, and the user must supply a wide range of input data in order to operate the model. The user-supplied data include a number of parameters for each component model used in the simulation, which characterize the size, type, in some cases thermal performance, and configuration of each and the way in which these components are interconnected. Each component is appropriately linked to other components to model the desired system.

A procedure developed at the University of Wisconsin which can be used for the design and economic evaluation of SHACOB systems is the f-chart (Klein 1976). Inputs to the interactive FORTRAN program include meteorological data, solar energy system data, building heat load specification, and economic variables. The required meteorological data are the long-term monthly average of daily total solar radiation on a horizontal surface, the long-term monthly average ambient temperature, and the long-term monthly average heating degree-days. These data are stored in the computer program for more than 250 stations in North America so that the user can select data which are representative of the test site.

The f-chart approach was developed because of the need for simplified design procedure to be used by architects, engineers, and builders. It is not strictly a simulation tool, since the thermal performance of individual systems under a set of weather conditions is not simulated. Rather, f-charts are design considerations based on hundreds of simulations of various system and building designs performed by Klein, using TRNSYS, at the University of Wisconsin.

Two dimensionless variables of solar heating systems have been identified as the important correlates of long-term system performance. The first is the ratio of the total energy absorbed on the collector-plate surface to the total heating load over some specified period. The second variable is the ratio of the maximum collector-plate energy losses (for water-based systems) to the total heating load during the period. The reference period is normally one month.

The factor f, the proportion of the total load which is supplied by the solar energy system, is related to the two dimensionless variables by means of a third-order polynomial regression equation. This relationship is then the basis for calculating monthly contribution to the total load of the solar energy system as a function of the input data for the building, the system, and weather.

The monthly f values can be derived from graphic representation or from the interactive computer model. These f values are based on simulations over a range of design parameters for Madison, Wisconsin, for one year of actual Madison weather. Klein (1976), however, suggests that the f values are relevant to long-term performance of a solar energy system such that they reflect not just one year of weather, but the historic record of Madison weather. In addition, Klein presents evidence that the f-charts are location-independent and therefore apply to any of the weather stations in North America.

The *f*-chart procedure provides not only an estimate of the contribution of the solar energy system to the total load but also economic analysis for the specified system. Data used for the economic analysis include the period of analysis in years, solar collector costs, fixed system costs, downpayment, interest rate on mortgage, term of mortgage, market discount rate, first-year expense, fuel cost, increase in fuel costs, and tax data. The economic analysis includes a routine to optimize collector size based on life-cycle costs. The model utilizes a numerical search to identify the collector size which maximizes discounted life-cycle savings. The analysis also includes a calculation of the number of years until positive savings occur and until cumulative discounted savings equal the mortgage principal (Kohler et al. 1977).

Verification of the Models

The component simulation models just discussed can model the thermal performance of SHACOB in very small time steps, depending on user specification. Therefore it is possible to map in detail the operation of the system *or* any one of its components to ensure the validity of the process based on known thermodynamic and architectural principles. In fact, the models are generally based on these principles expressed as mathematical formulations.

Adherence to the laws of thermodynamics, however, does not ensure that the simulation models accurately predict the long-term performance of a SHACOB system. Klein (1976) points out that if the models are accurate over a small time step (say, 1 h), then they should also provide accurate integrated quantities. This argument is appealing, but empirical evidence of the validity of the models in predicting long-term performance is needed.

There has been little or nothing published on the verification of some of the simulation models. (One major exception has been the work of Byron Winn at Colorado State University, Fort Collins.) The greatest amount of verification exists for TRNSYS, but only for short periods (say, 1 day) [see, for example, Klein (1976) Cooper, Klein, and Dixon (1975)]. Again we must apply the logic that if the model is accurate for short periods, it will provide acceptable long-term predictive values. But we know that even in a short period there are some prediction errors which may be compounded in the long run. It is also difficult to dismiss the variability and periodicity in weather as a contributing factor to long-run prediction errors. The problem of what constitutes a significant prediction error is important here, since the economic performance of the SHACOB system is dependent on the actual percentage of load supplied by the system. Therefore, a discrepancy between actual and simulated performance of, say, only 25 percent could have substantial impacts on the economic feasibilty of SHACOB.

The f-chart design procedure, which is based on the results of hundreds of TRNSYS simulation runs, has been verified by Klein (1976). He reports that when f is compared to actual values of solar contribution to total load for three solar-heated buildings for which experimental data exist, the f and experimental values correspond relatively well. He points out, however, that for Colorado State University (CSU) Solar House I the correspondence between the estimated and experimental values is poor in months with low heating loads, and the correspondence is good in months with high heating loads. The f-chart procedure is well verified with respect to TRNSYS, such that Klein estimates that the error magnitude between a value generated by f-chart and one generated by TRNSYS is smaller than the errors inherent in TRNSYS. Unfortunately, little is known of the validity of the other models, including SIMSHAC and SOLCOST, because of limited verification.

There have been some cooperative attempts to make comparisons between simulation models. At a conference sponsored by the Department of Energy (DOE) in 1977, the system simulation and economic analysis working group compared the results of solar space heating and water heating runs for f-chart and SOLCOST (Science Applications Inc. 1978). Comparable results for the annual solar fraction were obtained when the same solar insolation and load data were used.

A separate effort was made to compare TRNSYS and SIMSHAC results of the simulation of the CSU Solar House I, for which considerable performance data are available. These results showed that *both* TRNSYS and SIMSHAC accurately predicted the performance of the solar heating system.

It is our opinion, however, that further verification of the long-term predictive capabilities of, say, TRNSYS and the f-chart procedure is warranted. These two methods are extensively used by researchers in the solar energy field, and public policy with respect to SHACOB is being based in part on the results derived from these models. Certainly one way in which these models could be verified would be via an aggressive SHACOB construction and monitoring program. Such a program, however, would diminish the need for reliance on the results of simulation models, although these models would still be used in devising alternative designs and control strategies and in assessing the efficacy of new technology.

One attempt to circumvent the need for simulation models was made by Feldman and Anderson (1975b). Because of the inadequate verification of available simulation models, they attempted to develop predictive models of SHACOB system performance from regression analysis of actual solar building performance data. Their purpose was to examine the impact of SHACOB auxiliary electrical loads on electric utility peak-load capacity. It was therefore necessary to model the system performance on an hourly basis. The dependent variable used was storage tank temperature; the independent variables were lagged weather and system performance parameters.

They were able to develop reasonably strong predictive models, but abandoned the effort in favor of using TRNSYS to simulate auxiliary loads for three reasons. First, the building performance data which were available had large gaps and in other ways were not suitable for regression analysis. Second, many of the solar buildings for which detailed performance data are being collected do not reflect predicted uses of SHACOB. To illustrate this point, the CSU Solar House I, although designed as a single-family residence, is being used as a research laboratory. Building loads and system performance naturally reflect the actual, not the intended, building use. Third, it was not known what the effects on the accuracy of the predictive model would be when the model was exposed to weather for a different location or to weather beyond the bounds used to generate the model.

Criticism of the Models

Aside from the problem of verification, there are several other areas of criticism of the simulation models. One of these areas is the tradeoff involved between the simplicity and ease of use of the model and its accuracy. For example, some of the models use the relatively simple degree-day concept in the calculation of heat loads of the building, while the more complex models, such as TRNSYS, utilize extremely sophisticated algorithms to model the thermal performance of walls, the roof of the building, and even windows. Considerable knowledge of thermodynamics and formidable programming skills are needed to use these sophisticated models. It is also more expensive to run a simulation with detailed thermodynamic modeling, although direct cost comparison between models is difficult. So far the impact of using simplifying assumptions has not been adequately demonstrated.

A second area of concern involves the use of weather data as the time-dependent forcing function of the model. Since one of the primary objectives of the simulation is to predict the long-term performance of the SHACOB system, the model must be exposed to the character of long-term weather. This may be accomplished by running a simulation with many years of weather, such that some of the extremes in the weather as well as the long-term average are reflected in the data. This is not normally feasible, however, because each simulation run for many years would be very expensive and because for many locations there exists only a relatively short period of solar insolation data. For these reasons, most researchers have chosen to use only one year of weather data to predict long-term performance. In some cases, the year which has been used conforms to some conception of "average" weather; but for some locations, the limiting factor has been the availability of the insolation data, and so no choice of years was possible.

Klein (1976) has developed a promising approach to this problem. He has developed what he calls a "design" year by condensing nine years of Madison, Wisconsin, weather data into one. The method used was to examine all nine years of data on a month-by-month basis and to choose for the design year that month in the historic record which most nearly approximated the nine-year average for several key weather parameters. He tested the legitimacy of the design year by comparing results of simulations using the design year against simulations for which the entire historic record was used. He concluded that the design year adequately simulated the long-term performance of SHACOB systems.

There is, however, some question as to whether a design year or similar concept can actually reflect long-term performance. It may be, for example, that the design year does not adequately reflect long-term performance for all simulations of SHACOB systems. In addition, any averaging technique such as that used to produce the design year may tend to exclude the extremes in weather which may have significant impacts on system operation.

The variety of simulation models recently developed suggest that the technique has been accepted as a viable procedure for evaluating the performance and cost-effectiveness of solar buildings. The major advantage of simulation is that many alternative solar designs can be tested with varying weather data so that the performance of a specific system can be monitored. Thus, sensitivity analysis can be performed which could not be done by monitoring actual buildings.

Perhaps the primary limitation of solar simulation models is the lack of verification of the simulated results for the long term. The dearth of cases where actual building performance has been compared to simulated performance suggests that a certain degree of faith must be placed in the results of the simulation models. Also, the transient models which exhibit a high degree of isomorphism with actual systems are costly to run and require a sophisticated grasp of thermodynamics. These models require trained personnel to design the computer simulations and evaluate the results. While the less complex models are operated more easily, they cannot be used to simulate the detailed thermal performance of SHACOB systems. The simulation models which perform economic analyses suffer from a lack of standardization of terms. The economic jargon can be confusing for the model user. Simulation models are, nonetheless, powerful tools in the design of cost-effective SHACOB systems.

Building Model Specification

A much greater level of uncertainty exists not about the accuracy of the performance models themselves but about the specification of parameters.

Deviations in actual building practices cause a wide distribution in actual building performances. While the performance model is fairly capable of accurately predicting building performance given any single building practice, it remains a more difficult task to select the building construction practice and the related parameters that constitute the appropriate building model for this book.

In conforming to our objectives, every effort was made to specify parameters which favor a convenient null hypothesis, to ensure that no over-assumptive advantages exist for energy reduction techniques, and to see that energy-conservation investment is not favorable to active solar investment. When some question arose as to the specification of the parameter, every effort was made to select a value that was least favorable to investment in energy reduction. The model used here is TRNSYS, which is substantially modified for this book. In the most important instance, the infiltration rate for the buildings is not well defined within the TRNSYS simulation routine. Selection of infiltration rates—which normally range from one-half air changes per hour upward—was designated at the lower bound for all buildings. [See Peterson and Peterson (1979) for problems in calculating infiltration rates.] This underestimated energy demands and therefore made energy-reduction techniques less favorable. More importantly, the same low infiltration rate for all building designs tended to bias the less insulated building over the well-insulated and masonry structures because less insulated buildings normally experience more infiltration. The same process of biasing energy-reduction investments was utilized in the selection of all building parameters.

As was pointed out previously, the unwanted variability that exists in the building performance also conforms to this arrangement. The greater the energy-reduction investment, the higher the mean temperature within the deadband and the greater the unaccounted benefits of this additional investment. In addition, when comparisons between active solar energy and energy conservation were made, biasing toward active solar was attempted. The performance of the active systems approached the upper theoretical bounds; the costs and mechanical difficulties neared the lowest conceivable future conditions. As a result of these parameter specifications, some assurances as to the proof of the hypotheses are exacted.

The major aspect of the operational building model is the model of the structural building. This structure consists of the walls, roof, and room and determines the actual heating and cooling demand of the building. The actual specification of these building shell parameters greatly affects the building electricity consumption. Among the more critical of these parameters are the size of the building, orientation, insulation level, building construction weight, thermal mass, and infiltration.

One building parameter that affects the amount of electricity consumed is the size of the home. Size is important because some economies of scale in efficiency may exist for larger HVAC equipment. In addition, fixed costs are spread over a larger consumption unit. The home used in this study is 1,500 ft². The cost-effectiveness of some alternative technologies may be different if a much larger home was selected. However, economies of scale for conventional HVAC systems are also realized in larger homes, so it is unclear just how important size of the structure remains.

Orientation of the home refers to the position of the building's major axis. It is generally accepted that in the Northern Hemisphere a southern exposure for the major axis is desirable (except for perhaps the lower latitudes where cooling is the dominant concern). Some examination of the effect of orientations other than south on the electricity consumption and particularly peak demands is warranted. Variations in electricity demand also result from variations on the physical properties of the structure, particularly insulation level, construction weight, and infiltration.

One of the weaknesses of most reports on the economics of solar energy is the specification of insulation levels. No variation in building insulation is specified despite a large variation in actual insulation practices. The amount of insulation specified in these reports is usually below present preferred levels and has never been optimized with respect to other building parameters. For this examination, a range of insulation values has been identified, from no wall insulation to nearly 7 in of wall insulation. Attic insulation levels have also varied from 6 to 12 in.

The construction of the walls also affects the internal heat flow because of variations in construction weight. Three different construction weights have been specified by ASHRAE to distribute the energy flows among conduction, radiation, and convection. The designation of a building construction weight of light, medium, and heavy conforms to 30, 60, and 135 lb, respectively, of material per square foot of surface area.

The internal flow of energy is also affected by the heat capacitance within the internal living space. The value for capacitance helps specify the thermal mass of the building and includes interior portions, furniture, rugs, and so on.

The infiltration rate (in air changes per hour) is both an unclear and sensitive value. This is partly because of its diffuse nature, which makes monitoring difficult, and partly because of the importance attached to it in the TRNSYS algorithm for calculating convection flows within the building.

The existence of so many building structure variables necessitates the creation of a range of building types to limit calculations. Four different wall construction types have been utilized:

1. Base building
2. Energy-conservation building

3. Masonry building
4. Minimally insulated building

See table 2–1.

Table 2–1
Building Structure

Base Building

Type: Single-story ranch house
Dimensions: 58 ft long × 26 ft wide × 8 ft high (walls)
Volume: 12,000 ft³
Floor area: 1,500 ft²
Orientation: Long axis east-west
Basement: Yes
Attached structures: None
Wall construction: 2 × 4 in insulated stud wall; R11
Ceiling insulation: 6–in insulation; R19
Thermal mass: Light, 30 lb/ft² of building material
Effective capacitance: 13,920 kJ/°C
Window area: 16% of south-facing wall
              7.7% of east- and west-facing walls
              10% of north-facing wall
Glazings: Two panes
Infiltration: ½ air change per hour [a]

Energy-Conservation Building

Type: Single-story ranch house
Dimensions: 58 ft long × 26 ft wide × 8 ft high (walls)
Volume: 12,000 ft³
Floor area: 1,500 ft²
Orientation: Long axis east-west
Basement: Yes
Attached structures: None
Wall construction: 2 × 6 in insulated stud wall with 1–in total wall (Styrofoam insulation) on
                    outside; R24
Ceiling insulation: 12–in insulation; R38
Thermal mass: Light, 30 lb/ft² of building material
Effective capacitance: 13,920 kJ/°C
Window area: 16% of south-facing wall
              7.7% of east- and west-facing walls
              10% of north-facing wall
Glazings: Two panes
Infiltration: ½ air change per hour

Masonry Building

Type: Single-story ranch house
Dimensions: 58 ft long × 26 ft wide × 8 ft high (walls)
Volume: 12,000 ft³
Floor area: 1,500 ft²

Table 2–1 continued.

Orientation: Long axis east-west
Basement: Yes
Attached structures: None
Wall construction: 8–in masonry with 4–in total wall (Styrofoam insulation) on outside; R20
Ceiling insulation: 12–in insulation; R38
Thermal mass: Heavy, 130 lb/ft² of building material
Effective capacitance: 40,000 kJ/°C
Window area: 16% or 40% of south-facing wall
                     7.7% of east- and west-facing walls
                     10% of north-facing wall
Glazings: Two panes
Infiltration: ½ air change per hour

Uninsulated Wall Building

Type: Single-story ranch house
Dimensions: 58 ft long × 26 ft wide × 8 ft high (walls)
Volume: 12,000 ft³
Floor area: 1,500 ft²
Orientation: Long axis east-west
Basement: Yes
Attached structures: None
Wall construction: 2 × 4 in uninsulated stud wall; R4
Ceiling insulation: 6–in insulation; R19
Thermal mass: Light, 30 lb/ft² of building material
Effective capacitance: 13,920 kJ/°C
Window area: 16% of south-facing wall
                     7.7% of east- and west-facing walls
                     10% of north-facing wall
Glazings: Two panes
Infiltration: ½ air change per hour

---

[a]The infiltration rate utilized in this book represents an extremely well-constructed structure. This is an effort to minimize the importance of infiltration because of both the lack of sophistication in the TRNSYS model for infiltration and the uniqueness of infiltration and weather-stripping within each specific building. The designation of such a low infiltration rate tends to lower the cost-effectiveness of energy-reducing devices and practices.

Base Building. A base building has been specified in order to provide a basis of comparison for subsequent building modifications. The base building is a single-story ranch house with 2 × 4 in insulated stud wall (R11) and 6 in of insulation in the roof (R19). The building does not contain any other significant energy-conservation, active solar, or passive solar additions. The benefits of improvements in these areas can then be evaluated by comparison with this base building. This building conforms very closely to the new building conservation standards. All new residential construction must have this amount of insulation.

Energy-Conservation Building. Minimum building standards in no way represent the optimal amount of insulation and are not necessarily the most cost-effective alternative to the homeowner. A building designed for energy conservation was modeled to determine if such improvements are justified in Colorado Springs.

In the energy-conservation building, the 2 × 4 in stud wall is replaced by a fully insulated 2 × 6 in stud wall. As a second option, a 1-in sheet of total wall (Styrofoam) is nailed to the exterior. The application of Styrofoam not only increases the R value of the wall to 24, but also reduces conduction losses through the studs themselves. The use of 12 in of ceiling insulation increases the ceiling R value to 38.

Masonry Building. Materials other than wood are often used for building construction in Colorado Springs. Buildings utilizing masonry have different energy-consumption patterns than do wood buildings, because large amounts of building thermal mass lower the temperature fluctuations. At the same time, these materials have been used in buildings which utilize solar radiation and fenestration to control interior temperatures. These elements of building design, labeled direct-gain passive solar, have been shown to be an effective means of utilizing solar energy through building design.

A heavy-thermal-mass passive building has been modeled by using TRNSYS with modifications outlined in Abrash et al. (1978). The building construction consists of an 8-in masonry wall with 4 in of total wall sheeting on the external surface, resulting in an R value of 20. It is important that the insulation be put on the outside of the masonry so as to allow the heat stored in the masonry to be reradiated into the living space. The heat transmission properties (UA) of the passive building are very similar to those of the energy-conservation building. The difference in the performance of the two buildings can be attributed to fenestration and building thermal mass.

Minimally Insulated Building. This minimally insulated building represents the upper bound in building stock built prior to the 1970s. This building has 2 × 4 in uninsulated stud walls (R4) and 6 in of insulation (R19) in the ceiling. This category of building construction is included in an effort to explore retrofitting of both energy-conservation and solar energy systems.

Each of these buildings has different wall characteristics with regard to air resistance, heat capacitance, and energy transmission. The alleged difference in energy consumption and peak demand requirement of these four generic building types is of prime concern.

Other sources of energy are inputted to the room. The use of lights and appliances generates heat energy which must be distributed to the room. People themselves generate heat, although the occupancy of a building is variable. The sensitivity of human occupancy on heat loads of buildings is an area of little knowledge. The variation of use habits is so great that virtually no usable performance data can be gleaned from the Department of

Housing and Urban Development (HUD) solar demonstration projects now in operation.[12] Because of this problem, this model is forced to hold the human occupancy constant, despite the potential importance that this variable has on building loads.

Within these larger categories of building type, several other construction practices have been noted as important in energy consumption. The sizing and design of fenestration are important determinants of building energy demands.

Fenestration. Specifications within the model include window orientation, window size, number of glazings, size of overhanging shade, and use of thermal shutters. Since typical homes have little concern over the placement of windows, to test the sensitivity of window size and placement, two different window scenarios were established. The standard window scenario had 16 percent of the south surface in glass and 10 percent of the north surface in glass. In order to examine the feasibility of solar direct gain through windows, 80 percent of the south wall was made glass and the north glass surface was reduced to 3.5 percent. Variations in these scenarios were performed to test the sensitivity of orientation and number of glazings.

The second aspect of fenestration deals with overhanging shades and shutter routines. For many of the scenarios a 4–ft fixed overhanging shade was added to the south roof structure to limit solar heat gain during the cooling period. Experimentation with a removable shade such as an awning or vegetation-covered trellis was not performed, but would be desirable because removable shades do not restrict winter solar heat gain.

Insulated shutters were modeled to lower heat losses at night during the winter and direct gain in the summer. The biggest modeling problem of these shutters was the development of realistic control strategies. Mechanical controls exist which could operate shutter closings based on solar radiation levels or time of day. These controls are inflexible and cause occasional inefficient operation, particularly in the spring and fall. This problem normally would be corrected by human intervention. In the same context, one of the simplest ways to cool a home is through opening windows. Modeling difficulties preclude the actual implementation of this logic, but a count of the number of times that open-window cooling potential existed is recorded.

The Heating, Ventilation, and Air Conditioning Equipment. All the homes utilize typical HVAC equipment to maintain comfort levels, including electric auxiliary heating, air conditioning, domestic hot-water systems, and control devices. Table 2–2 is a description of the systems used in this book. The auxiliary heating system is modeled as a forced-air system because of the ease of incorporation with active solar heating.

Air conditioning is provided similarly from a 3–ton electric compression air conditioner with a relatively high coefficient of performance (COP) of 2.4. More accurate modeling of air conditioners is warranted in summer-peaking utilities where the COP should vary with temperature. The third component of the HVAC equipment is a domestic hot-water heater modeled with a preheat tank. The all-electric domestic hot-water system ignores losses from the preheat tank so that it operates similarly to a single-tank system. The use pattern for hot-water demand is developed from Mutch (1974).

The smooth operation of a building's space conditioning system depends on several control devices to regulate temperature. Control devices such as thermostats and time clocks can reduce energy use and obtain greater comfort control. All the controls utilized here are modeled with logic routines that are commercially available in mechanical and microprocessing devices.

In reality it is possible that human behavior can substitute for these control devices. The operation of the building's heating and cooling system is then more dependent on individual whim and discipline. The logic of the human being is normally the same as that of the machine, but the ease and inexpensiveness of modern thermostats would suggest that some of these devices are strikingly cost-effective and time-saving.

The operation of the room space-conditioning system requires a set of controls to command heat from the solar tank or auxiliary electric heater and to cool from the electric air conditioning. The operation of these control devices utilizes a three-stage controller to regulate the first three requirements.

The three-stage controller operates as any normal heating and cooling thermostat would except that a third mode is added to account for solar heat. The heating element has two modes. If the temperature is below 20°C, then heat from the solar storage tank will be circulated. When the tank heat is insufficient and the temperature drops below 18.3°C, the auxiliary electric heater is activated, adding 36,000 kJ to the room every time step. (The gap between the two heating bands allows the room temperature to vary. This creates some problems in comparisons of different simulation where these average room temperatures differ.) The upper band of the three-stage controller, 25.0°C, activates the auxiliary compression air conditioner with a capacity of 38,400 kJ.

In all cases the temperatures used as minimum and maximum room temperature before heating and cooling by auxiliary are the government-suggested thermostat settings. While these values may be extreme for most houses, the cost-effectiveness of this customer action exceeds any benefits from any other available behavior or technical energy-conservation option. From a microeconomic perspective, modeling of buildings with less extreme thermostat settings is a deviation from the optimal. For those individuals unwilling to alter thermostat settings, this analysis represents a worst case. For them, the cost-effectiveness of energy conservation and solar energy investment will be more pronounced.

The Active Solar System. The modeling of a solar collector system is also accomplished by using TRNSYS. The sun's energy is absorbed by the collector and transferred to storage via a pump. The loop also contains a relief valve and thermostatic control systems to maintain operation. The collected energy is transferred to the storage tank via a heat exchanger. Energy from the tank is inputed to the tank as required when the tank temperature is sufficient. The simulation closely resembles an actual collection system in operational logic.

The collector system described in this simulation is of the flat-plate variety. Flat-plate collectors represent commercially available technology. The collector principally performs a three-step process. The collector allows solar radiation to penetrate the glass plates and be absorbed by the black nonreflective surface. Some of the heat energy is lost to the environment through the top and bottom of the collector. Finally, some of the heat energy absorbed in the back plate is transferred to the fluid passed over the plate.

In reality, the collector performance is extremely erratic and dynamic, being dependent on an ambient temperature, collector temperature, and fluid temperature flow rate. A bigger physical modeling problem lies in accounting for the change in fluid temperature and thus collector performance as the fluid passes through the collector.

Duffie and Beckman (1974) have outlined a method for generalization of collector performance for the wide variety of temperatures instantaneously within the collector. Briefly, the generalized collector model requires three parameters that are most critical to performance: the transmittance-absorption product $\tau\alpha$, which determines the percentage of solar radiation collected; the heat-loss coefficient UL, which determines the amount of heat losses of the collector; and the flow rate FR, which calculates the amount of useful energy collected off of the collector. Together these three parameters account for the greatest variability in simulation results of collectors. These values and other collector parameters are listed in table 2–2. In this simulation very high efficiencies have been used, which represent an extremely well-operating collector.

Several other parameters concerning collectors are emphasized in much of the literature; these include the tilt and orientation of the collector. The determination of optimal tilt and orientation angles is dependent on not only the available solar radiation but also cost comparisons and collection procedures. It is normally assumed that a southern orientation is optimal to ensure maximum insolation values. Some deviation from true south may be

Table 2-2
Solar and Space-Conditioning Systems

Space-Conditioning Systems

Active Solar System

Collector area: 35 m^2, 17.5 m^2, 70 m^2 (combined system with hot water)
Collector tilt: 60° or 45°
Number of glass covers: Two panes
Collector efficiency factor F':   0.9
Collector-plate absorptance α: 0.9
Collector-plate emittance Σ_p: 0.9
Fluid thermal capacitance C_p: 3.56 kg/(kg · °C)
Loss coefficient for bottom and edge U_{be} :   1.63
Extinction coefficient × thickness of each glass cover K_L: 0.037
Flow rate: 879.0 kg/hr
Heat exchanger mode: Counterflow
Overall heat-transfer coefficient of exchanger: 11,870
Tank volume: 1.32 m^3, 2.56 m^2, 5.12 m^2, 0.88 m^2, 7.68 m^2
Loss coefficient for tank U: 1.021
Pump on/off temperatures: 5.6°C/1.7°C
Auxiliary heating system: Electric forced air
Air-conditioning system: 3–ton conventional compression-cycle air conditioner; COP 2.4
Interior thermostat: Three-stage
   Minimum utilized solar storage temperature: 25.55°C
   "Cooling on" temperature: T interior > 23.33°C or 25°C
   "Solar heating on" temperature: T interior < 20.00°C
   "Auxiliary heating on" temperature: T interior < 18.4°C

Passive Solar System

Shading: Horizontal overhang shade on south wall
Shutters: Insulating shutters on windows; for window and shutter combined
Energy control routines: Count kept of number of times windows could have been opened to
   provide needed cooling; shutters opened at night in summer for radiative cooling and closed
   during day to reduce direct heat gain; shutters closed at night in winter to reduce heat loss
   and opened during day to maximize direct heat gain.
Interior heat generation
   Energy transfer into room from people: 1,100 kJ/h
   Energy transfer into room from appliances and lighting fixtures: 1,800 kJ/h

---

Domestic Hot-Water System

Collector area: 5.6 or 11.2 m^2
Collector tilt: 60°
Number of glass covers: Two
Collector efficiency factor F':   0.9
Collector-plate absorptance α: 0.9
Collector-plate emittance Σ_p: 0.9
Fluid thermal capacitance Cp: 3.56
Loss coefficient for bottom and edge U_{be} :   1.63 kJ/(h · °C · m^2)
Extinction coefficient × thickness of each glass cover K_L: 0.037
Flow rate: 281.0 kg/h
Pump on/off temperatures: 5.6°C/1.7°C

Table 2–2 continued.

Domestic-Hot-Water System

Effectiveness of heat exchanger: 0.7
Volume of preheat storage tank: 0.303 m³
Loss coefficient of preheat tank U: 1.80
Use Profile: Typical daily mass flow profile
Total daily mass demand: 300 l
Delivery temperature: 48.9°C

more optimal if one part of the sun's path is usually blocked by clouds, buildings, or trees.

More concern is generated over the proper collector tilt. The optimal tilt angle varies depending on the use of the collector. Summer collection requires a lesser tilt from horizontal. Determination of the optimal tilt will be changed if the rationale for collection changes from best average collection performance to best collection performance during extreme conditions. It may be that optimization of collector tilt in a winter-peaking utility with a marginal-cost pricing tariff may be nearer vertical in order to collect radiation best during winter-peak conditions.

The real issue in regard to optimization of collector tilt is the cost consideration.[13] A constraint to maximizing solar radiation potential becomes the cost of constructing a roof with a pitch of nonstandard size. Standard building practices usually require a roof pitch of 0°, 30°, 45°, 60°, or 90°. It is these values for collector tilt that are being tested here.

At present, other collector configurations are not commercially available for low-temperature technologies. Mention of both the evacuative collector and the concentrating collector is warranted by their potential for much higher collection temperatures, thus eliminating some of the constraints to solar air conditioning. The cost of these collectors, however, and the insufficient testing of reliability make modeling of their performance at this time impractical.

The performance of the flat-plate collector is dependent on the flow rate of the transfer fluid. A pump forces the movement of this fluid through the system. The pump model behaves similarly to an actual pump. If the pump is on, the fluid is circulated at the maximum pump rate. There is, of course, no fluid movement when the pump is not operating. In these simulations, the flow rate of the pump is proportional to the collector size. Pump electrical energy is added to the total electricity as 3.5 percent of the total solar energy collected.

The operation of the pump is commanded by a thermostatic controller which evaluates the temperatures of the collector and the storage tank. If the collector temperature is below a minimum temperature or less than that of the storage tank, then operation of the pump would cause a net loss in storage temperature. This mode of operation may be acceptable for radiational cooling, but certainly not for solar heating. The thermostatic model compares the collector temperature to the storage temperature and initiates operation when there is a 5.6°C temperature difference. The pump continues to operate until the lower band temperature difference between collector and tank (1.76°C) is reached.

Temperatures of water collected from flat-plate collectors are normally less than 100°C. It is possible on hot, sunny summer days to exceed the 100°C temperature and thus cause the liquid to boil in the tank. In this instance, the system will release pressure by means of a pressure relief valve. TRNSYS incorporates a relief valve algorithm, which merely removes energy from the loop until the temperature is below the maximum allowable.

From the collector, the heated fluid is directed to the storage facility. The possible storage media for solar heating systems have been limited to liquid, rock bed, or change of phase. For our purposes a water storage tank is utilized because of its compatibility with liquid collection systems.

The optimal sizing of storage systems is a particular unresearched area of solar buildings. Traditionally, rule-of-thumb figures for the sizing of storage capacity always have linked the storage size to a proportion of the collector size. This process stems from the fact that the storage process directly affects the collection process, so that optimization of the storage system separately was unfeasible.

The basic problem (beyond cost considerations) with the sizing of a storage system is the effect of storage size on collector efficiency. The performance of the collector decreases as its temperature increases. Therefore, system efficiencies are higher when the collector is capable of operating at low temperatures or, in other words, when the storage-tank heat-transfer-fluid temperatures are low. To lower the tank temperature without reducing the heat content of the storage, it is necessary to increase the volume of storage. A larger tank can retain the same amount of energy as a smaller tank, but the larger tank will have a lower temperature. Juxtaposed to this action is the cost consideration that larger tanks impose. In addition, the heat-loss figures increase with an increase in surface area of the tank.

The storage tank's prime purpose is to provide adequate heat energy to the load, both space conditioning and hot water. In this respect, the temperature of the tank has very little effect on the capability of the storage to meet loads. The critical value is the heat content of the storage. A larger tank at a lower temperature can provide the required heat to load just as well as a smaller tank with the same heat content but a higher temperature.

By its very nature, storage must concern itself with providing energy to load when the collection system is least capable of providing that energy.

The sizing of storage is then affected most by the tradeoff between the costs of larger storage and the costs of auxiliary sources. The past practice of sizing the collector-storage components for maximization of total annual collection may be in conflict with the *true optimum*. When it no longer can be assumed that auxiliary energy is available or inexpensive, then the role of storage becomes more critical.

The storage system can behave as a storage mechanism for auxiliary off-peak energy just as for the solar collection system. The reduction of on-peak demands can be accomplished by immersing the auxiliary heat source within the tank and supplementing the tank with electric resistance energy during off-peak periods. The efficacy of this procedure is reduced by the unnecessary waste of auxiliary energy. The storage system will be heated on nights prior to good solar-collection days, and the rise in storage tank temperatures will lower collector efficiencies.

The possibility of maintaining a larger tank or even two tanks, one for solar and one for off-peak auxiliary, has been suggested. [One example of the use of two storage tanks is the Goosebrook solar home that utilizes a separate off-peak unit as well as the solar storage unit. See Shurcliff (1978).] It is probably more likely that the auxiliary off-peak storage system alone is presently more cost-effective, thus eliminating the need for the solar equipment altogether.

This simulation examines several storage operation modes. The storage tank utilized in the simulation is a cylindrical tank that is well insulated with a loss coefficient of 1.02 kJ/(h · m³ · °C). It is located outside the heated space and modeled with a constant ambient temperature of 19°C in order to simplify computation of active and passive heating gains.[14] The height of the storage tank is 2.134 m; the diameter and thus the volume are varied to model different collector/storage ratios.

The TRNSYS model provides for tank stratification to better model the heat flow in water tanks. The degree of accuracy in that model was not felt to be critical enough to warrant the additional running expense. The unstratified tank model therefore assumes a constant mixing of water throughout the tank.

The auxiliary energy is never supplied to the tank. The heating and hot-water loads are met by an exiting pipe from the tank. If the heat energy in the tank is insufficient, then auxiliary is supplied directly to the room or the preheat hot-water tank. In this manner, the auxiliary energy does not interfere with the active collection of solar radiation.

The delivery of heat to the room is accomplished by another pump similar to the collector-loop pump. The key to the operation of this system is the three-stage controller. The circulation of heat from the tank to the building loads begins whenever the temperature drops below 20°C and heat is required. At 18.4°C the auxiliary energy augments the solar system.

Passive Solar Contribution. As mentioned previously, the passive solar contribution is calculated by using improvements in the TRNSYS model. The energy that enters through the windows is distributed by using lag transfer coefficients described earlier. These lag coefficients are set by the user depending on the construction weight specified in the room model. The designation of a building as medium construction is equivalent to 60 lb of water per square foot. The heavy construction weight of 135 lb/ft^2 is achievable in the heavy masonry buildings only.

There is a problem with the use of TRNSYS in modeling passive buildings in that TRNSYS treats solar heat gain through windows in a perfunctory fashion. Transmission through glass is specified as a constant; the solar radiation on each surface is reduced by a constant shading function; no provision is made for differential glazing between sides; no shutter routine is available; and the treatment of diffuse radiation is ignored.

A routine to account for solar radiation through each window and the calculation of an external shading device were developed here and described in Abrash et al. (1978). This routine enables the accurate determination of contribution of windows to direct gain. Another addition, thermal shutters, allows for the inclusion of this important technique in passive design.

The direct-gain technique of solar passive design is the least sophisticated of passive techniques. Its deficiency is that the solar radiation through the windows is distributed throughout the room. Therefore, the passive description presented cannot accurately assess the effectiveness of a Trombe wall partition or even water tanks that are directly in the path of the sunlight. The model is, in effect, assessing a suboptimal configuration where thermal mass is distributed throughout the room.

Determination of Building Cost

The determination of economic efficiency for the solar building studied, from consumer, utility, and national welfare perspectives, relies on the calculation of building impact on energy use and demand and the cost of the building configuration producing that impact. This subsection discusses the methodology used for costing specific building components and presents incremental cost data for the building configurations studied.

The selection of the particular construction types represents a balance between actual construction practice for conventional single-family residential buildings and the project team's expectations of what construction practices may be for mass-produced solar and energy-conservation single-family residential buildings. For example, the base (2 × 4 in stud wall) building reflects standard industry practice and was modeled and costed accordingly. However, in selecting a heavy-mass, well-insulated building,

Table 2–3
Building Descriptions Used in the Case Study Simulation

To be concise, there are four building types which are generic:

Base building: 2 × 4 in insulated stud wall, 6-in attic insulation
Energy-conservation building: 2 × 6 in insulated stud wall, 12-in attic insulation
Masonry building: 8-in concrete block, 4-in exterior Styrofoam, 12-in attic insulation
Minimally insulated building: 2 × 4 in uninsulated wall, 6-in attic insulation

There are also three generic HVAC system types:

A: Combined 35-m² solar heating and hot water, electric backup and cooling
B: 5.6-m² solar hot-water system, electric heating, cooling, and auxiliary
C: All-electric heating, cooling, and hot water

In the Colorado Springs case study sixteen buildings were modeled with each HVAC system type:

Building number:

1: Base bulding; 16% south windows
2: Base building with overhanging shade; 16% south windows
3: Energy-conservation building with shade; 16% south windows
4: Energy-conservation building with 1-in Styrofoam, with shade; 16% south windows
5: Energy-conservation building with 1-in Styrofoam, 80% south windows, with shade
6: Energy-conservation building with Styrofoam, 80% south windows, thermal shutters, with shade
7: Energy-conservation building with triple storage size in active system 7.95 m³
8: Energy-conservation building with one-third storage size in active system 0.88 m³
9: Energy-conservation building with 17.5-m² collector, 1.32-m³ storage, with shade
10: Energy-conservation building with 70-m² collector, 5.30-m³ storage, with shade
11: Masonry building with shade; 16% south windows
12: Masonry building, 80% south windows, with shade
13: Masonry building, 80% south windows, thermal shutters, with shade
14: Masonry building, 80% south windows, winter/summer shutters, no shade
15: Masonry building, 80% south windows, triple glazing, with shade
16: Minimally insulated building with shade

Therefore, as an illustration, building number 4A is the energy-conservation building with 1-in Styrofoam, with a shade and 16% southern window area. This building's HVAC system is a combined 35-m² solar heating and hot-water system with electric backup and cooling.

no standard industry practice exists for residential buildings—especially a standard practice that would typify industry practice ten to fifteen years from now when such buildings may impact on utility loads. Therefore, it was necessary to select specific construction details that balanced the desired thermal performance characteristics, realistic construction practice, and installation costs. While this balancing of objectives is addressed in the discussion of each building component, it is best exemplified by the selection of the high-mass building envelope. In providing a high-thermal-mass building, it was decided to use masonry exterior-wall construction. The final selection of solid concrete block represents a balance between thermal performance, where poured-in-place or precast concrete may be best, and residential building practice, where conventional nonsolid concrete block may represent the current industry approach in conventional buildings.

For final cost determination for each building component option, the published architectural cost handbook *Means Construction Cost Data* was employed. It is important to understand that the costs available from this handbook represent costs applicable to larger commercial structures and not specifically to residential buildings. While labor rates are for union trades and therefore represent labor costs higher than what would be expected for residential construction, the material costs are for large commercial projects where volume buying allows a lower material cost than would be available to the typical residential builder. Thus the final cost determination is believed to be representative of residential construction practice, but is subject to considerable variation for a specific component. The variation in building component cost is somewhat moderated by the selection of building components that, in balancing thermal performance and residential building practice, usually result in a moderate cost option, and not necessarily in an option at either end of the cost range.

Costing Methodology. The costing procedure utilized here is similar to that used by many architects in arriving at preliminary cost estimates. The primary difference is that the methodology allows for the cost determination of insulated building components in order to facilitate the aggregation of these components into a number of different scenarios.

The various tasks associated with the costing methodology are:

1. Disaggregate buildings into components to be costed.
2. Select an appropriate architectural treatment for each component.
3. Prepare architectural drawings and specifications in sufficient detail to cost components.
4. Determine component costs and, where appropriate, determine incremental costs over specified base cost.
5. Adjust component costs for selected cities.
6. Aggregate building component costs for specified buildings.

Tasks 1 to 3, associated with costing methodology, are discussed in conjunction with the following subsection on building components. The four primary energy-conservation and solar energy parameters (insulation, thermal mass, fenestration treatment, and active solar energy system designs) are characterized by specific building components. Where there exists a range of architectural options for a building component which all accomplish the same objective (say, three ways of providing window shading are covered), one option is selected based on thermal performance, standard construction practice, probable consumer acceptance, and cost. The selected building-component options are then characterized by architectural drawings and specifications in sufficient detail to permit costing.

The building component costs are determined through the use of the published construction-cost guide *Means Construction Cost Data*. In order to have a consistent cost base with the utility data, all construction costs are reported in 1976 dollars, using the 1976 edition of *Means*. As discussed in the opening paragraphs of this chapter, *Means* is specifically developed for commercial construction and not directly applicable to the residential construction considered here. In any case, the costs are determined on a consistent basis, and while the reported costs may not be directly applicable to residential construction, the relative costs are believed accurate.

Component costs calculated with the use of *Means* include material, labor, overhead, and profit on a standardized basis. These standardized costs are then adjusted for specific localities by using city cost indices in *Means*. These cost indices reflect the relative building cost differences for specific locations in the United States.

Once all component costs are determined, overall building designs are specified. The specified buildings are costed by aggregating the proper building components and summing the costs. Incremental costs of specific buildings over the cost of the base case then serve as inputs to the financial analysis from consumer, utility, and national-welfare perspectives. Actual building costs are aggregated as part of the financial analysis.

Building Components. The building components selected for study reflect those components that might be included as part of energy-conservation or solar heating packages. An attempt is made to exclude or hold constant those components which would not be a direct part of such packages. (See table 2–1.) For example, while solar buildings, especially passive solar buildings, may have nonconventional floor plans to facilitate natural air circulation, the selected buildings have the same floor plan—one that is amenable to conventional and solar building design alike. Interior construction (partitions, conventional plumbing, appliances, furnishings, and the like) is assumed constant. Building components beyond the building proper (garages, patios, vegetation) are excluded. While both interior structure and components beyond the building proper offer significant opportunity for impacting on the consumer, utilities, and national welfare, they are beyond the scope of this book.

The four parameters most affecting auxiliary energy consumption are insulation level, building thermal mass, fenestration treatment, and active solar energy system sizes. Insulation and building thermal mass are characterized by building envelope specification. Fenestration treatment involves window orientation and area, window shading and shuttering devices, and the number of glazing layers. The active solar energy system is specified as a "conventional" liquid system of various collector sizes and collector area–storage volume ratios. Roof structures are modified as necessary to accommodate varying collector sizes. In that there is significant overlap in building cost elements among the four energy conservation–solar energy parameters, it is necessary to individually cost the building elements affected by each parameter and then aggregate the building elements so as to represent a specific combination of energy conservation–solar energy parameters. The building elements selected for costing are ceiling insulation, window overhang (window shading device), roof structure, wall structure including insulation and fenestration, fenestration treatment, and active solar energy system costs.

Ceiling Insulation. All buildings modeled have a minimum of 15 cm of fiber glass ceiling insulation. It is recognized that a large fraction of new and a larger fraction of existing single-family dwellings do not have the equivalent of 15 cm of fiber glass ceiling insulation (R19). However, the 1985–1990 period should see the adoption of R19 ceiling insulation as an absolute minimum; therefore, 15 cm was selected as the base cost.

The ceiling insulation was increased for more energy-conserving buildings to a level of 30 cm. Ceiling insulation levels in the range of 15 to 30 cm may be changed without materially changing other building elements.

Window Overhang. In passive solar buildings, a stationary projection or overhang is placed above windows to allow the low-altitude winter sun to enter but to block the high summer sun. Such window shades may be provided by attached horizontal trellises, offset roofs, or ceiling joist extensions. Any of the three approaches to providing window shading may be selected depending on the architectural requirements of the building. A 1.42-m overhang, rather than the present typical overhang of 0.36 m, provides the desired shading effect. All three approaches were costed, and the approach using ceiling joist extensions was selected because of its low cost and its minimal impact on other architectural elements (such as roof structure).

Roof Structure. The roof structure for a roof-mounted solar energy system differs from a conventional roof. This difference results from the need for the collector to be at some optimal angle for collecting the winter sun. A standard roof was assumed to have a minimum 4/12 pitch (to accommodate snow buildup in some study locations), to be built from prefabricated roof trusses (typical of production ranch-style homes), and to have asphalt roofing. This roof structure was modified to accept space-heating collectors (17.5 m² and 35 m²) by maintaining the 4/12 north roof pitch but increasing the south pitch to 19/12 (60°). In order to accommodate the larger 70-m^2 collector, the south roof surface was increased, which increases the slope (and surface area) of the north roof. The incremental costs of the modified roof structures were calculated as the additional cost of lumber for larger roof trusses, additional sheathing, building paper, trim and rake edges for the roof, and the additional area of the end walls. Incremental costs for asphalt roofing, excluding shingles from under collectors, were calculated. The standard roof structure was used for the 5.6-m^2 solar domestic hot-water system by providing for a support frame in the cost of that system.

Wall Structure. Costs for the various wall structures were complicated by the interdependence of structural type (2 × 4 in, 2 × 6 in, masonry), insulation amount and type, and fenestration considerations.

Seven structural/insulation configurations were initially selected for costing:

2 × 4 in stud wall without insulation

2 × 4 in stud wall with R11 insulation

2 × 4 in stud wall, R11 insulation, 1–in Styrofoam

2 × 6 in stud wall, R19 insulation

2 × 6 in stud wall, R19 insulation, 1–in Styrofoam

8–in solid concrete block with 2–in externally applied Styrofoam

8–in solid concrete block with 4–in externally applied Styrofoam

The 2 × 4 in stud wall, with studs 16 in on center is typical of the vast majority of residential construction. The 2 × 6 in stud wall with studs 24 in on center is gaining rapid acceptance as an incrementally low-cost method for allowing thicker insulation to be installed in wall cavities. Styrofoam is gaining rapid acceptance as an insulating, nonstructural sheathing and was considered as an addition to both wood-frame wall constructions. As discussed earlier, the selection of solid concrete block as the structural component for a high-mass wall is based on balancing thermal performance, standard construction practice, and cost. In both masonry wall scenarios, insulation (2 in and 4 in) is applied to the outside surface of the masonry.

After initial costing, it was possible to eliminate two wall types—the 2 × 4 in stud wall with R11 insulation and 1–in Styrofoam and the 8–in solid concrete block with 2–in Styrofoam. The first was eliminated because it offered slightly less insulating value at a higher cost than the 2 × 6 in stud wall with R19 insulation. Although a number of builders now offer the 2 × 4 in stud wall with total wall construction, it is assumed that buildings with this construction will become less frequent as more builders become aware of the benefit of 2 × 6 in stud walls. The 8–in concrete block with 2–in Styrofoam was eliminated because of minor cost differences with the 8–in concrete block with 4–in Styrofoam and the fact that it offered significantly less thermal performance.

Three fenestration scenarios were applied to all wall types. East- and west-wall glass areas were held constant at 8 percent for all cases. South and north window areas were varied as follows: 16 percent south glass, 10 percent north glass; 40 percent south glass, 4 percent north glass; 80 percent south glass, 4 percent north glass. The 16 percent south glass is typical of ranch-style homes with a sliding glass door. The 40 and 80 percent south glass cases provide significantly more window for passive gains; therefore, the north-side windows were reduced to reflect a desire to minimize losses out those windows. Standard window sizes were used throughout.

Since increasing the glass area reduces the wall area, costs for the three window configurations and the five different wall types were done as a matrix. Costs of the windows were figured to include interior trim, flashing, caulking, and exterior painting, as well as the cost of the window units themselves. The remaining area of the wall for each elevation was multiplied by a cost per unit area calculated for each wall type. The masonry wall costs also include an additional cost per linear foot of window opening for concrete lintels and sills. The cost for the 80 percent glass, south-wall elevation was calculated once and kept constant because the remaining 20 percent consists of framing and trim and so the cost did not vary with wall type.

Fenestration Treatment. With the heat loss/gain through glass representing such a large percentage of the building load for the 80 percent glass south elevation cases, two energy-conservation scenarios were applied to the windows. The base case window system utilized double-pane wood-frame windows. The first energy-conservation scenario utilizes a third layer of glass. This cost was calculated on a unit-area basis and applied to all windows. The second window energy-conservation scenario involves the selection of insulating shutters and an appropriate operating scheme. Three shutter options, all providing R11 insulating capability, were designed and costed: Styrofoam window inserts; multilayer, insulating roller shades; and sliding, solid shutters. The multilayer, insulating roller shades were selected based on moderate cost and probable consumer acceptance.

Although specific operating routines were used in modeling the performance of the shutters, they were assumed to be manually operated, and no costs were included for automation.

Active Solar Energy Systems. The collector system selected is a typical liquid, flat-plate system. Collector areas and storage volume are varied under various scenarios. Costs were derived per unit collector area, per unit storage volume, and for fixed costs, so as to give reasonable cost when aggregated into a system. These costs are $160 per square meter of collector, $256 per cubic meter of storage, and $1,500 for fixed aspects of components such as pumps, piping, and controllers. For the base 35-m^2 system, the cost is $7,842, or $224 per square meter.

The installed cost of a solar domestic hot-water system, including a collector support frame for the roof, was assumed a constant $2,500 over a conventional electric hot-water system.

Operating and Maintenance Expenses. In addition to installed capital costs for the various solar buildings, operating and maintenance expenses play an important role in the economic analysis. Operating costs for the active collector systems are primarily pump running costs. Pump energy is assumed to be 3.5 percent of the collected energy and is included in the cost analysis along with the auxiliary energy consumption. Incremental maintenance expenses were calculated for all solar building designs. However, the only significant incremental maintenance costs were found to be for the collector system, window overhang, and shutters.

After modeling the solar building in its thermal performance and financial costs, we then proceed to compute utility costs according to criteria which are consonant with these calculations.

Modeling the Utility

In most studies of the economics of alternative energy sources, the utility company's production function is virtually ignored. In effect, the electric bill under present rate structures is considered to be the extent of required information concerning the utility. This thesis has suggested the inadequacy of this approach to the utility, the policy analyst, and the building owner. The costs to the utility of providing service to the solar building are in no way reflected in the average cost rates presently implemented in most utilities. Concomitantly, the building owner cannot rely on these rates to provide him with an indication of the cost of auxiliary electricity for his system. A more sophisticated model of the electric utility must be presented to arrive at these answers. The model must include simulation of utility loads, the relationship between loads and costs, and the relationship between these costs and tariffs.

Earlier we identified a procedure for calculating the diversified electric consumption (hour by hour) of a residence. The impact of these residences on utility costs and profits remains to be calculated. The load of this new building is treated as an increment to the existing load of the utility, and the cost of this increment in load is determined. The cost to the utility is compared to the revenue received by the utility from the building (the electric bill) with regard to major imbalances between costs and revenues.

The economic principle by which the cost to the utility should be evaluated is the marginal cost of production, the cost of a given change in loss-of-load probability. This value represents the true opportunity cost to the utility of increments (or decrements) in load. Several problems emerge, however, in moving from the theoretical construct of marginal costs to operationalizing the definition and the calculations of these costs. We touch upon the production accounting basis for operational approaches here; this issue is discussed in more depth later.

Calculating Marginal Costs. One debate among utility economists is the relevant time perspective for determining marginal costs. The literature specifies the theoretical importance of using the short-run marginal costs (SRMC).[15] The example of a streetcar illustrates a problem with using a short-run marginal-cost perspective. As long as the streetcar has empty seats, the marginal cost of an additional occupant is low because only slightly more energy is needed to power the machine and there is little wear and tear on the cushion. The marginal cost of each subsequent passenger remains low until the unlucky individual who requires an additional seat when there is none. A longer run perspective would consider the cost of a new streetcar before the last seat is occupied.

Most long-run marginal-cost methodologies represent an economic approximation of the optimal future investment plan. In order to calculate this curve, several assumptions have to be made concerning the future demand and cost of electricity. The first step is the specification of a demand schedule for electricity. This is a pragmatic problem given the lack of knowledge with regard to elasticity of demand for electricity.

The next step is to determine a production function which in some way represents an optimal growth scenario. How the methodology determines this production function is dependent on many assumptions about electricity growth, the time pattern of analysis, future costs of generation capacity and fuel, plant availability, security constraints, investment criteria, and historic utility composition. Turvey (1968, p. 59) warns against the theoretical attempts to calculate long-run marginal costs:

> The same fact of technical progress makes long-run marginal cost a much more complicated concept than any economics textbook allows. There is no escaping an element of judgment in its calculation. It has been shown that this judgment is required concerning a large number of variables but it can also be packed into one portmanteau guesstimate about depreciation.

> Finally, it cannot be emphasized too strongly that any estimate of long-term marginal cost has no significance "in abstracts" but only in relation to a specified load increment. There are as many marginal costs as there are conceivable load increments.

Given that there is no clear-cut evidence to adopt any one particular approach, we attempt to satisfy approximations of long- and short-run marginal cost using an adaptation of the Cicchetti-Gillen-Smolensky (1978) model, which is relatively easy to manipulate computationally.

Calculating Energy-Related Costs. The basis for this short-run decision framework is "system lambda." System lambda represents the engineering dispatch routine by which the utility optimizes its generation mix. These short-run decisions are based on the lowest available fuel-cost alternative for meeting existing load. The same information regarding dispatching provides a good indication of marginal costs as they vary with variations in load. The primary component of short-run marginal costs is energy or fuel costs. The demand in kilowatts of an electric utility is met at any time by a group of generating plants with sufficient capacity. Each of the units making up this capacity has a fuel requirement per kilowatt hour with regard to thermal efficiency. The cost of that fuel, be it gas, oil, coal, or nuclear, is the incremental fuel cost of that plant. See figure 2–1. (It should be noted that this figure may vary with the output of the particular plant in question, but this variation is not as significant as plant-to-plant variation.) The utility operates a system of economic dispatch where those plants with the lowest incremental fuel costs make up needed capacity.

A good measure of marginal energy-related costs is the incremental fuel cost of the last plant to be put on-line. In figure 2–1 the system lambda for a utility has been generalized into four types of generating facilities.

In the example in figure 2–1, at 3:00 a.m. the new fossil-fuel plant with an incremental cost of 8 mils/kWh would represent the marginal cost. At 6:00 p.m. the addition of almost 300 mW of load has pushed the marginal generation plant to the gas peaking turbines, with an incremental cost of 30 mils/kWh.

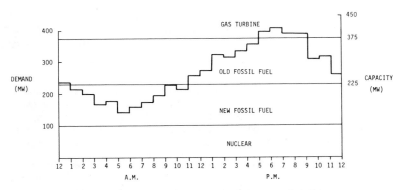

Figure 2–1. Load Profile and Generation Types

Some utilities have incorporated a load model such as this one to provide accurate, hour-by-hour, energy cost data. The real value of this model lies in its ability to assess the cost of increments in load such as the costs of an incremental solar building. An example of this type of model is shown in Feldman and Anderson (1976a).

The big utility models can distinguish cost differences between each variation in load. Generalization of plant efficiencies into broad categories still provides a good representation of running costs. In the examples utilized here, these generalized efficiencies have been designated to conform with the specified demand periods of the utility. The least efficient generation plant normally used during each period signals the incremental fuel cost.

Calculating Marginal Capacity Costs. The energy-related cost model explained above suffers from the same deficiency noted as a problem in theoretical short-run marginal costs. When the demand for electricity exceeds 450 mW, as is the case in figure 2-1, the short-run cost curve approaches infinity.

Turvey suggests a method of accounting for the capacity constraint problems that uses a short-run cost calculation which negates any theoretical differences between short- and long-run marginal costs. He redefines the short-run pricing rule so that "when demand exceeds capacity, price is set at the level necessary to restrict demand to capacity output" (Turvey 1968, p. 92). In the calculation of marginal generation costs, a procedure is utilized similar to that suggested by Turvey. Since large generation facilities require so much lead time, future construction plans are fixed for the most part.

Unlike the theoretical long-run marginal-cost curve, utilities cannot optimize and reoptimize their production plans to conform to changes in growth of electricity demand by altering the size of future generation facilities. The real alternative available to the utility planner is to accelerate or defer the construction timetable of a fully conceived facility to make up for deviations from anticipated growth. To fill this gap in supply, each utility also has the option of purchasing electricity from other utilities or purchasing small gas-peaking facilities. The costs of these three options signify the marginal generation costs.

In order to represent the real marginal generation costs to the utility, the costs figures obtained from the above accounts must be adjusted to include power losses and reserve-margin requirements. Losses in power through transmission and distribution mean that the kilowatts required at the place of end use must be translated into additional capacity at the point of generation. The need to maintain reserve capacity also increases the amount of generation capacity maintained by the utility.

Cicchetti, Gillen, and Smolensky (1976) realize the complexity of cost accounting for electric utilities. To help simplify the procedure, sensitivity analysis is performed on most cost data to determine the required accuracy. Major margins of variation in cost signify the need for separate cost accounts. For this reason transmission and distribution costs, including capital expenditures, and loss multipliers should be specified for each voltage level.

The Cicchetti-Gillen-Smolensky method of calculating marginal costs is more convenient than some of the other methods. Most of the required data is easily available to electric utility planners in such a manner as to provide as accurate an account of marginal costs as is necessary. The greatest benefit beyond simplicity is the consistency of data across utilities. This provides comparisons of marginal costs from utility to utility.

If we know the marginal cost of generation capacity, we then must calculate the increment in capacity required as a direct result of the new building load. Both Lorsch (1977) and Asbury and Mueller (1977) assumed that the capacity requirements of solar buildings and of all-electric buildings would be equivalent. A methodology, described previously in this chapter, was developed herein to quantify this relationship.

The procedure is to calculate the probable building demand in kilowatts at the time of system peak by extrapolating from simulated building performance during recent utility peaks. By using the modified TRNSYS, each building is simulated for five days prior to each of the past five annual peak-load days. The performance of the building on the peak day of the utility is obtained. The desired result of the simulation is the diversified electric demand for the peak period (the hours for which there is probability of loss of load) on the peak-load day. The demand is then diversified over time, since electric draw for space conditioning and water heating occurs in short bursts. The model uses the average demand of the building during the peak period of the peak-load day for each of five years as a surrogate for capacity requirement. It is not useful to specify the demand of the building during the instant of system peak because of the lack of precision in the building model and in utility forecasting.

The use of historic peak weather circumvents the problem of trying to use utility forecasting weather-load models to generate ample peak weather conditions. The lack of knowledge about both weather sequences and the probability distribution of extreme weather events makes the use of actual weather conditions appear superior. The real problem with the use of historic data is that the weather-sensitive loads of many utilities have undergone significant changes because of higher saturations of space-conditioning equipment. The further one goes back, the less relevant the weather ditions associated with that peak occurrence are to future peak occurrences.

The original methodology (Feldman and Anderson 1976a) utilized ten historic years of weather. Because the methodology averages the ten-year performances, equal weight was given to each of the years. The present method for calculating capacity requirements has reduced the number of historic years to five. The choice of five years is almost arbitrary. It is understood that weather conditions historically tell little about the weather conditions to be expected next year at peak; yet given the distribution of extreme weather events, it is necessary to utilize more than a single previous year's weather as a surrogate.

A further problem with the above methodology may be the placement of capacity charges only on the annual peak load. Utilities may experience near-peak conditions on several other days so that some capacity burden extends beyond the one system peak occurrence. Complicating this issue is the fact that electricity growth may cause shifts in system peak to new seasons or leapfrogs from season to season. Variations in supply through hydropower availability or maintenance scheduling may also cause capacity constraints without experiencing system peak.

The entire issue of capacity requirement, however, reduces to a question of reliability and forecasting of capacity requirements—whether electric utilities can depend upon solar energy building as capacity-reducing or whether full backup capacity must be provided. These issues require answers, particularly with regard to larger utilities, where service areas may extend between several climatic areas, and in pricing decisions, where cost of production is finely scrutinized. This is the topic of the next chapter.

Notes

1. This review of literature considers those reports that examine directly the microeconomic feasibility of solar energy. Another body of literature is the market penetration studies. These reports use predictions as to the cost-effectiveness of solar energy to project the contribution of solar energy toward meeting future energy needs. These reports most often do not utilize detailed analyses of tradeoffs between different solar technologies, but estimate growth of solar energy based on trend analysis, often relying on more detailed methodologies as the basis of these projections. These market penetration reports which heavily influence federal support for solar technology are subject to the same inconsistencies that are found in the more detailed solar economic studies. Among the more often quoted include Bennington and Rebibo (1976), Stanford Research Institute (1977), and Arthur D. Little, Inc. (1977a). Also see Schiffel, Costello, and Posner (1978) for a review of state-of-the-art market penetration methodology.

2. See also Löf and Tybout (1973b, 1974).

3. See also Roach et al. (1977) and Schulze et al. (1977).

4. The reason that solar feasibility is not ensured merely by being competitive with electricity is illustrated by the assertion that car A which

gets 4 mi/gal would be a good investment considering that existing car B only gets 2 mi/gal. While car A is better than car B, it is certainly not the best investment. Optimization of a car's efficiency obtains from treating the car as a total system. Consideration of the engine efficiency may not produce as good results as consideration of engine, weight, tires, and so on taken all together. All the previously mentioned reports treat the solar system as the optimizing function. Because solar air-conditioning technology is not commercially viable, the study centers only on the heating load; however, benefits of conservation extend through the cooling load as well. It is important to consider the entire space-conditioning and hot-water load in order to optimize the consumer's investment.

5. The New Mexico Study used a methodology outlined in Balcomb and Hedstrom (1976).

6. In 1975 ASHRAE 90–75 was established as a new building minimum standard for residential building energy conservation. It requires the equivalent of a fully insulated 2 × 4 in stud wall and 6–in attic insulation. ASHRAE 90–75 was established as a minimum standard, and therefore opportunity may exist to include much larger amounts of energy conservation; see ASHRAE (1975).

7. See Cone et al. (1978) or Fassbender and Cone (1978).

8. The total ΔT for a period is represented by the degree-day or degree-hour concept, where $\Sigma \Delta T = \Sigma(65°F - T)$. For a detailed discussion of calculating UA and degree-days, see Anderson (1977) or ASHRAE (1977).

9. Cal-ERDA has recently been renamed DOE-2 (Graven and Hirsch 1977); Hisper (Nasa-Marshall Space Flight Center) and LASL (Los Alamos Scientific Laboratory) are not readily accessible to the public.

10. The model PASOLE (McFarland 1978) has been developed by Los Alamos Scientific Laboratory and is also described in Balcomb and McFarland (1978). The procedure is to specify nodes from which energy transfer proceeds. A simplified procedure for hand-held calculators was developed along these lines by Kohler and Sullivan (1978); see also Heldt (1978) for other passive modeling techniques.

11. The modifications to TRNSYS for passive solar modeling include the use of shutters, overhanging shades, and more accurate accounting of direct solar heat gains. See also Abrash et al. (1978). A new version of TRNSYS, VERSION 10, incorporates similar components.

12. Clearly other factors have resulted in the lack of usable data from the HUD demonstration—principally the original intent to demonstrate rather than research solar energy technology. Better-designed research projects are needed to determine consumer use patterns. The present studies by Princeton University (Socolow et al. 1978) and Electric Power Research Institute (Arthur D. Little, Inc. 1977b) are important first steps.

13. Anderson (1977, p. 201) shows that in Minneapolis variation in collector size because of different tilt angles is small. The difference between a 50° tilt and a 60° tilt is about 6 percent. The effects when one moves south are even smaller. This is confirmed by Feldman et al. (1979) for collector tilts of 45° and 60° in California.

14. The placement of the tank outside the heated space is to allow comparisons of solar-heated buildings with all-electric buildings, where the all-electric version is the sum of the solar energy supplied to the room and the electric auxiliary used by the solar building. This allows the modeling of both buildings with a single TRNSYS run. Having the tank external to the room overestimates heating loads, but underestimates cooling loads for the solar building; but it was felt to be worthwhile given the expense of each TRNSYS run and the unknown aspects of storage heat-loss modeling.

15. Theoretically, marginal costs are calculated as infinitesimal increments in load in the short run, where all capital is fixed, because the short-run marginal cost "reflects the social opportunity cost of providing the additional unit that buyers are at any given time trying to decide whether to buy" (Kahn 1970, vol. 1, p. 71).

3

Solar Energy and Electric-Utility Sales and Forecasting

Douglas Woods

The forecasting of future electricity demand is important to the effective management of most business firms, but it is particularly vital for the electric utilities. Electricity sales forecasts are necessary for cash budgeting and planning fuel requirements, while short-term peak-load forecasts are essential for determining the need for additional peaking capacity. Long-range demand projections are required to plan additions to base-load capacity.

In determining their capacity requirements, the electric utilities find themselves in something of a dilemma. A large portion of the total cost of supplying electrical power to their final users is accounted for by the fixed cost of capital equipment employed in generation, transmission, and distribution. In order to minimize these costs the utilities would like to keep their reserve margins as small as possible by keeping the planned level of capacity close to anticipated peak demand; however, to guarantee the availability of power when needed, they must plan for substantial margins of reserve capacity. The only way to reconcile these two objectives effectively is through the accurate forecasting of future demand.

This chapter examines the impact that the growth of residential solar space conditioning (heating and cooling) and water heating will have on electric-utility load forecasting. While the primary focus of the chapter is on the effect of the use of solar energy in housing units rather than commercial establishments, the methodology discussed here could be readily extended to a consideration of the latter, with some modification, of course. The methodology could also be employed to analyze the effect of the widespread introduction of solar water heating alone on electric utility loads, a development which, in the eyes of some observers, may precede the introduction on a large scale of solar space conditioning.

The chapter surveys some of the more popular methods available for forecasting both electricity consumption and peak demand and comments on their effectiveness. It examines those techniques that have actually gained widespread acceptance by the power industry and the newer methods that are being developed. It then considers how these techniques may be adapted or extended to take into account the probable impact of an increase in the use of solar energy in the future.

The relationship between solar's impact on expected or average peak demand and the above-normal extreme peaks which determine utility capacity requirements is examined. The chapter focuses on the problems of forecasting at the level of the individual electric utility rather than of the United States as a whole or large regions of the country. Since the rate of introduction of solar space conditioning and water heating is likely to vary substantially, depending on the region of the country examined, the discussion necessarily tends to emphasize methodology rather than actual projections of the rate of growth of solar energy and its impacts on utility loads.

Solar Energy and Utility Loads

It is obvious that the use of solar space heating and cooling and water heating will reduce a residence's consumption of other forms of energy. But it will not necessarily reduce the demand for electricity. In many cases solar space heating is likely to be substituted for some type of heating system other than electrical. Where this occurs, electrical consumption may increase. The solar space-heating system will require electric power to circulate the heat exchange medium through rooftop collectors to the thermal storage reservoir, in addition to the normal consumption of electricity to circulate hot water or hot air throughout the building. Moreover, as mentioned previously, the auxiliary source of energy in a solar space-heating system is likely to be electric resistance heating.

Since the demand for auxiliary energy is likely to be relatively small, large initial investment costs will not be justified—making electricity the most economical choice. Clearly, then, the net effect of the widespread adoption of solar heating on electric energy consumption depends on whether typically it is adopted in place of gas, oil, or coal heating systems or in place of electric resistance heating or heat pumps. With respect to cooling, the situation is obviously somewhat different. In the residential sector space cooling is accomplished almost entirely through electrically powered compression air conditioners. Consequently, the growth of solar space cooling will unquestionably reduce the consumption of electricity in that application. The effect of solar water heating on electricity use in water heating depends on factors similar to those which will determine the net impact of solar space heating on electrical consumption. The direction of the effect turns on whether solar hot-water heating grows at the expense of electric water heating or largely at the expense of water heating by natural gas or oil.

Up to this point, the discussion of the impact of SHACOB on utility loads has focused entirely on the potential change in electric energy consumption rather than on peak demand. Unfortunately the effect of solar energy on future capacity requirements is even more ambiguous than its impact on energy consumption. The reason is not difficult to identify. The contribution of a solar space-heating system, for example, to a building's heating requirements, can decline during the same seasons and under the same weather conditions that increase the total demand for electric power— especially in winter-peaking systems.

Winter-peaking utilities normally experience their peaks during late December and early January in periods of very cold weather when the number of daylight hours is at a minimum. The limited daylight at these times obviously tends to minimize the output from solar space-heating systems. Moreover, the sun's low angle to the horizon reduces the intensity of radiation and the available solar energy. Fortunately, the peak winter periods for most utilities occur during extremely cold weather when skies are clear, thus affording considerable insolation and permitting solar space- and water-heating systems to make a significant contribution toward the utility system peak. Nevertheless, the reduction of a utility's winter peak brought about by the introduction of solar heating frequently may not be proportionate to the reduction in electric energy consumption.

It is clear that the widespread utilization of solar energy for space- and water-heating purposes has serious implications for the accuracy of power demand forecasting. Many uncertainties surround the rate at which solar energy systems will be adopted and the various forms which they may take. These uncertainties greatly complicate the problem which faces the electric utilities of accurately predicting the demand for power ten and twenty years hence.

At the same time, it is important to recognize that utility load forecasting will have a significant impact on the economics of solar energy. The savings in capacity costs which the use of solar space and water heating may make possible will actually materialize only if the utilities anticipate its happening. If they do not expect a substantial reduction in their future peak demands because of solar energy, they will plan their capacity expansions accordingly, and the materialization of significant solar energy use will have no effect on the total fixed cost of electric power generation. In short, if solar energy is to contribute to a reduction of the electric utilities' capacity costs, it must be anticipated. Moreover, because of the long lead times required for the installation of new capacity, it must be anticipated as much as ten years in advance.

Electricity Sales Forecasting: A Note on the Current State of the Art

Power demand forecasting methodology is not far advanced at present. Until very recently the techniques employed to predict both electricity sales and peak demand have been relatively unsophisticated. The electric utilities themselves have tended to rely heavily on extrapolative techniques. While there have been a large number of econometric studies of electric energy consumption, the published literature on peak demand has been very limited, indeed. Moreover, despite the substantial work done on the development of econometric models that can be used to project future electric energy consumption, these models have gained relatively limited acceptance by the utility industry forecasters. This section briefly reviews the state of the art of forecasting electric energy consumption. Its focus is on the econometric methods that have been developed and are now beginning to be applied in forecasting the residential sales of individual electric utility companies. A subsequent section examines the problem of taking an anticipated technological change, such as the introduction of solar energy, into account in an econometric load-forecasting model.

It is widely recognized among researchers in this field that the residential demand for electricity depends in the short run on the stock of energy-consuming appliances possessed by the household and on the rate of utilization of that stock. The latter is a function of the current cost of electricity and the household's present income level. In the long run the primary determinants of the household's demand for electricity are the factors influencing its purchases of appliances, including space-conditioning equipment and lights. These factors include the prices of electricity and appliances and the incomes of consumers as well as the prices of competing forms of energy. In many studies both the current and the lagged values of these variables appear as independent variables in the demand functions. Since a household's present stock of appliances has been built up gradually over a number of years, the stock is a function of past as well as current prices and income levels.

The techniques used to estimate the residential consumption equations have varied substantially. An excellent overview summary is provided by Aigner 1979. Several equations have been estimated by using annual time-series data drawn generally from the past two decades. (Houthakker and Taylor 1970). In some cases, quarterly data have been used. Other studies have been cross-sectional in nature, looking at how consumption has varied among different states or SMSAs at the same time in response to income and price differentials (Wilson 1971). Some studies have employed pooled cross-sectional and time-series data (Mount, Chapman, and Tyrrell 1973). In most residential electricity-demand equations, the dependent variable is electricity consumption per capita. Total consumption is obtained by multiplying the equation by the population.

In spite of the variation in estimation techniques, the studies have encountered many of the same problems. To begin with, there has been difficulty in handling the electricity rate variable, primarily because of the declining block-rate structure employed in the residential sector by most utilities. The use of average revenue (total residential revenue divided by total residential sales) has the disadvantage that a negative relationship between consumption and average rate results simply from the fact that as consumption grows over time, more and more households move into rate blocks in which the marginal electricity rate is lower, thus pulling down the average price which they pay. On the other hand, if a marginal rate is used (that is, the cost per additional kilowatt hour in one of the rate blocks), other problems arise. The most obvious is the difficulty of determining which rate block is the appropriate one to use. This problem is complicated by the fact that the marginal block of the average household has changed over time as households have increased their consumption and moved into lower rate blocks. Taylor (1975) suggested that the rate in the marginal block, the price per kilowatt hour which the household currently pays for additional amounts, and the average rate for all the intramarginal blocks combined be included as price variables in the equation for electricity consumption of the household. He reasons that while changes in the rate for the marginal block have both an income and a substitution effect on consumption, changes in the average rate for the intramarginal blocks have only an income effect. The expenditure each household has to make in order to purchase the total quantity of electricity in the intramarginal blocks and thus the household's residual income is changed, but the cost per additional kilowatt hour of energy consumed is not altered. However, the average household's annual expenditure on electricity is a small percentage of its annual disposable income. Thus, the impact on electricity consumption of changes over time in the average intramarginal rate will be swamped by changes in the household's total disposable income. As Taylor notes, a bias will be introduced by including only the marginal price in the demand equation; however, that bias should be quite small.

A second significant problem encountered with the econometric electric-energy consumption models estimated to date (at least as far as their usefulness to utility forecasters and the willingness of the latter to employ them is concerned) is the unrealistic estimate of the electricity rate and income elasticities generally produced by most of these models. The long-run elasticity of residential electricity consumption with respect to the electricity rate typically has been between -1 and 2. Income elasticities, on the other hand, have varied very widely, but generally have been less than 1 and in some cases have actually been negative. Most of the electric utility industry regards an electricity rate elasticity well above 1 as unrealistically high. Indeed, it is hard to reconcile high rate elasticities and low income inelasticities. Since electricity consumption in the long run is largely a function of the stock of appliances owned by households and since purchases of appliances are largely determined by consumer incomes, one would expect income to be a more important determinant of electricity consumption than the rate. The price elasticity of demand for electricity to operate appliances will tend to be high relative to the income elasticity when it is possible for consumers to substitute appliances using other types of energy for electricity-consuming appliances. This kind of substitution is, of course, possible for hot-water heating, ranges, and ovens. However, for most appliances, such as refrigerators, freezers, air conditioners, televisions, and lights, gas- or oil-operated alternatives either do not exist or are highly inferior. The elasticity of electricity demand with respect to the electricity rate will also tend to be high relative to the income elasticity, provided that consumers are willing to make substitutions between appliances in general and other consumer goods. A priori one would expect that their willingness to make such substitutions would be relatively limited.

Most studies have not included the price of appliances as a determinant of electricity consumption, and this is perhaps their third major failing. The present discounted cost of owning and operating many appliances over their useful life is dependent more on their initial cost than on the price of electricity. Of course, this circumstance tends to vary from appliance to appliance. For example, it is certainly true of color television sets but not hot-water heaters. With respect to heavy users of electricity like ranges, refrigerators, and freezers, the initial purchase price and the present discounted value of the electricity consumed are likely to be of roughly equal orders of magnitude.

The necessity of distinguishing between the short- and long-run demand for electricity is the fourth major problem encountered in the development of explanatory equations for household consumption of electricity. In dealing with this problem econometricians generally have adopted one of two alternative strategies. The first involves estimating separate equations for the short and long runs. For example, Fisher and Kaysen (1962) developed functions to explain short-term changes in the rate of utilization of appliance stocks. They also estimated explanatory equations for about half a dozen major appliances, relating appliances stocks to disposable incomes, electricity, and gas prices as well as a number of demographic variables such as the number of marriages and the ratio of residential and rural electric customers to the population. Anderson (1973) developed market-share equations for the major appliances. In these equations the ratio of each electrical appliance's share of the total market over the share possessed by a gas appliance or an appliance using heating oil is treated as the dependent variable, while gas and electricity prices and other important determinants of appliance purchases appear as the independent variables.

The other major method for solving the problem of distinguishing between the short- and long-run demand for electricity includes both the current and past values of the determinants of demand in the equations for electricity consumption. Generally this has been accomplished by incorporating into the right side of the equations the dependent variable itself lagged one period. This autoregressive functional form has the effect of making consumption depend on the current and all past values of the independent variables with declining weights on the lagged values going back in time. Unfortunately, autoregressive equations have important statistical and forecasting disadvantages. These can be avoided largely by the use of fixed-weight lag structures or Almon polynomial distributed lags.

The approach employed by Fisher and Kaysen, in which the long-run demand for electricity is equated to the demand for appliances and the short-run demand is determined by their rate of utilization, makes possible the convenient handling of the anticipated growth of solar power in developing load-forecasting equations. Under this procedure solar space conditioning and water heating can be treated simply as new types of appliances. Explanatory equations can be estimated in which the dependent variable is the ratio of the solar share of total sales of all space-conditioning and water-heating systems to the sales accounted for by other types of space-conditioning and water-heating systems. Unfortunately, this methodology is hard to apply to the development of energy forecasting models for total residential electrical consumption. One obvious difficulty is that although total electricity consumption in a household depends on its entire stock of appliances, including lights, typically individual equations cannot be estimated for all appliances because of a lack of data. Moreover, there exists no convenient means for taking into account changes over time in the amount of electricity consumed per appliance. Such changes are important and occur for a number of reasons. The efficiency of appliances varies in response to variations in fuel and electricity costs; in addition, the average size of appliances tends to increase. Nevertheless, many experts in the field believe that these problems can be overcome and that the disaggregated appliance-stock models represent the best prospect for improved forecasting of residential electricity sales.

Because of the complexity of disaggregate appliance-stock models, most econometricians and utility forecasters have developed aggregate forecasting equations for the residential sector. In the following discussion, it is assumed that an aggregative method is used to generate a base forecast of future electric energy consumption. The residential electricity demand equation will be assumed to have been estimated from time-series data and to contain the lagged as well as the current values of the independent variables. Ideally, the equation will include the marginal electricity rate, competing energy prices, an average appliance price, and disposable incomes, all deflated, and such demographic variables as population and the number of marriages.

The base forecast of electricity consumption is assumed to largely ignore the effect on future utility sales of an acceleration in the rate of growth of solar energy utilization. Therefore, a forecast of the impact of solar space conditioning and water heating has to be factored into the base forecast. The impact of solar energy is treated as an add-on, as a modification to a forecast that has been made without taking into account the accelerated growth of solar power.

Incorporating a Forecast of Solar-Market Penetration into the Forecast of a Utility's Future Electricity Sales

The development of SHACOB presents a special problem for electric-energy sales forecasting because a forecast based on data from the historical period cannot account adequately for an anticipated change in the rate of growth of solar energy use in the future. During the historic period SHACOB will, in many cases, have had a zero or virtually negligible impact on utility sales and peak demand. A base forecast produced by a model considering only the historic period will not properly predict the future growth of SHACOB. This occurs because the cost factors or the consumer attitudes underlying the homeowner's choice among alternative methods of space conditioning and water heating may change in the future so as to promote the greater utilization of solar energy. Nevertheless, a base forecast of sales, which excludes the effect of the potential growth of SHACOB, is the starting point for the calculation of future electric energy sales. To modify this base forecast, four types of information are required; each must be specific to the area served by the electric utility and must cover the entire forecast period:

1. An estimate of the average number of kilowatt hours of electricity required as auxiliary energy and to circulate the collection medium and operate the control devices in each type of SHACOB system
2. A forecast of the average annual electricity consumption of each type of nonsolar, conventional system
3. A forecast yielded by a study of possible capture of the solar market involving the number of new and old homes that will have SHACOB systems and systems using each of the competing energy sources for every year of the forecast period.
4. An econometric projection based on historical data of the number of new and old homes that would have had each of the alternative types of HVAC systems were it not for the accelerated growth in solar energy utilization

The difference between these two forecasts of the future stocks of all types of space-conditioning and water-heating systems—the one being based strictly on past relationships and the other taking account of an expected increase in solar energy use—will be combined with the estimates of the amount of electricity used by each type of HVAC system to modify the utility's base forecast of its residential sales.

The methodology of the solar market penetration impact study is discussed below. This section deals with the problem of obtaining projections, based on historical relationships, of the shares of the space-conditioning and water-heating market accounted for by each alternative energy source.

For this purpose the utility forecasters must eliminate relative market-share functions, similar to those developed by Anderson, in which the proportion of homes powered by an alternative system is treated as a function of such explanatory variables as the ratio of the initial investment costs of the two systems and the ratio of their annual operating costs. However, since an individual utility is necessarily concerned with demand in a single, limited geographical area, these equations are estimated over time-series rather than cross-sectional data. Estimating relative market shares rather than calculating stocks directly is appropriate when the decision faced by a homeowner is not whether to own an appliance but rather which one, as is the case with space-conditioning and water-heating systems. In most areas of the country, homes having no system constitute only a tiny fraction of the total. Anderson's estimation technique makes use of the fact that relative market shares must sum to 1.

Anderson demonstrates that each fuel's relative share of the total sales of all appliances of a given kind (for example, a HVAC system) is a function of relative fuel prices, appliance prices, income, and demographic variables. The dependent variables in his relative market-share equations are S_{it}/S_{mt} for $i = 1, ..., m - 1$, where i and m designate individual fuel types and S_{it} and S_{mt} are the mt shares of total sales of fuel types i and m, respectively, over the relevant period. In Anderson's study the period was from 1960 to 1970; consequently, he could assume that the sales of appliances using a given fuel type were equal simply to the total stocks of those appliances in use at the end of the decade.

In developing relative market-share equations from time-series data for a given area, the period over which sales are measured must be restricted to a year, not a decade. Consequently, sales are more closely related to the change in stocks during the year than their level at the end of the year, that is, to ΔH_{it}, $= H_{it} - H_{i(t-1)}$ rather than H_{it}, the stock of the appliance using fuel types in existence at the end of year t. Unfortunately, ΔH_{it} accounts for only a portion of the total sales of the HVAC systems using fuel type i in a single year.

Where Q_{it} = total sales of HVAC systems using fuel type i in year t
    d_{it} = the proportion of homes with HVAC systems using fuel type i
        that are demolished in year t
    r_{it} = the proportion of homes with HVAC systems using fuel type i
        that replace their existing system in year t

we have

$$Q_{it} = \Delta H_{it} + (d_{it} + r_{it})H_{i(t-1)} \qquad (3.1)$$

The product $(d_{it} + r_{it})H_{i(t-1)}$ is the number of homes using fuel type i at the beginning of year t that either are demolished or replace their existing system in year t. The total change in the stock of space-conditioning and water-heating systems of fuel type i in operation in year t, ΔH_{it}, is equal to the total sales of type i systems Q_{it} less the number of i-type systems that must be replaced in that year or are lost as a result of the demolition of homes. Thus, in addition to knowing ΔH_{it} and H_{it} for all $i = 1, ..., m$, which can be calculated largely from electric- and gas-utility records, it is also necessary to determine d_{it} and r_{it}, the rates of demolotion and replacement for all systems $i = 1, ..., m$ before the relative market-share equations can be estimated.

One suspects that in many cases r_{it} will have to be estimated from knowledge of the average lives of space-conditioning and water-heating systems and the proportion of these systems using each fuel type in the past. Provided that H_{it} can be determined for values of t several years prior to the beginning of the historic period over which the relative market-share equations are to be estimated, it will be possible to compute the age distribution of the stocks of HVAC water-heating systems of fuel type i in each year of the period of estimation. From the age distribution and an estimate of the probability distribution of the operating life of system type i, the expected value of r_{it}, the proportion of the units of i that need replacement in year t, can be computed. A similar process may be required to calculate d_{it}.

The equations for relative market shares are derived, following Anderson, from the assumed relationship

$$\frac{S_{it}}{S_{jt}} = \frac{g_{it}(\cdot)}{g_{jt}(\cdot)}$$

where $S_{it} = Q_{it}/\sum_{j=1}^{m}Q_{jt}$ is the share of the total sales of HVAC systems that use fuel type i and $g_{it}(\cdot) = (P_{it}, V_{it}, R_{it}, Y_{t}, Z_{t})$. In the functions $g_{it}(\cdot)$ for $i = 1, ..., m$, P is the price of fuel type i; V_{it} is the price of the system using fuel type i; $R_{it} - r_{it}H_{i(t-1)}/\sum_{j=1}^{m}r_{jt}H_{j(t-1)}$ is the proportion of the total number of existing systems that must be replaced in year t which use fuel type i; Y_t is personal disposable income per capita; and Z_t is a vector of other demographic variables. (These would include average family size, the proportion of total dwelling units that are single detached, and the proportion of nonurban housing units.) Assuming that $g_{it}(\cdot)$ can be written in the form $g_{it}(\cdot) = P_{it}^{b_{1i}} V_{it}^{b_{2i}} R_{it}^{b_{3i}} Y_t^{b_{4i}} Z_t^{b_{5i}}$ for all $i = 1, \ldots m$, we have

$$\frac{S_{it}}{S_{mt}} = \frac{a_{it}}{a_{mt}} \frac{P_{it}^{b_{1i}}}{P_{mt}^{b_{1m}}} \frac{V_{it}^{b_{2i}}}{V_{mt}^{b_{2m}}} \frac{R_{it}^{b_{3i}}}{R_{mt}^{b_{3m}}} Y_t^{b_{4i}-b_{4m}} Z_t^{b_{5i}-b_{5m}} \qquad i = 1, \ldots, m \qquad (3.2)$$

Equation 3.2 will be estimated from time-series data drawn from the historical period and will be substituted into equation 3.6 to forecast H_{it}, the stock of HVAC systems using fuel type i in the future year t, for all $i = 1, \ldots, m$.

Equation 3.6 is derived as follows. By definition of S_{it},

$$Q_{it} = \frac{S_{it}}{S_{mt}} S_{mt} \left(\sum_{j=1}^{m} Q_{jt} \right) \qquad i = 1, \ldots, m \qquad (3.3)$$

Since

$$S_{mt} = \frac{1}{1 + \sum_{i=1}^{m-1} (S_{it}/S_{mt})}$$

and

$$\sum_{j=1}^{m} Q_{jt} = \sum_{j=1}^{m} \left(H_{jt} - H_{j(t-1)} \right) + \sum_{j=1}^{m} \left(r_{jt} + d_{jt} \right) H_{j(t-1)}$$

then

$$Q_{it} = \frac{S_{it}}{S_{mt}} \frac{1}{1 + \sum_{i=1}^{m-1} (S_{it}/S_{mt})} \left(\sum_{j=1}^{m} H_{jt} - H_{j(t-1)} \right) +$$

$$+ \sum_{j=1}^{m} (r_{jt} + d_{jt}) H_{j(t-1)} \qquad i = 1, \ldots, m \qquad (3.4)$$

Since $Q_{it} = \Delta H_{it} + (r_{it} + d_{it}) H_{i(t-1)}$,

$$Q_{it} = H_{it} + (r_{it} + d_{it} - 1) H_{i(t-1)} \qquad (3.5)$$

Substituting equation 3.4 into 3.5 and rearranging terms give

$$\frac{S_{it}}{S_{mt}} \frac{1}{1 + \sum\limits_{j=1}^{m-1} (S_{it}/S_{mt})_{ij}} \left[\sum\limits_{\substack{j=1 \\ ij}}^{m} H_{jt} + \sum\limits_{\substack{j=1 \\ ij}}^{m} (r_{\cdot jt} + d_{\cdot jt} - 1) H_{\cdot j(t-1)}) \right] +$$

$$\frac{S_{it}}{S_{mt}} \left[\frac{1}{1 + \sum\limits_{i=1}^{m=1} (S_{it}/S_{mt})} - 1 \right] \left[H_{it} + (r_{it} + d_{rt} - 1) H_{i(t-1)} \right] = 0$$

$$i = 1, ..., m \hspace{4cm} (3.6)$$

Once equation 3.2 has been estimated and r_{it} and d_{it} for $i = 1, ..., m$ forecasted, all the variables in equation 3.6 are predetermined except H_{it} , ..., H_{mt} . (The future values of d_{it} and r_{it} for all space-conditioning or water-heating system fuel types can be forecast by the same methods that were employed to estimate the values of d_{it} and r_{it} during the historic period.) When their values are substituted into equation 3.6, a set of m linear simultaneous equations is obtained which can be solved by Cramer's rule for the values of the m variaables H_{it} , ..., H_{mt} .

The second step in modifying the base forecast of the utility's electricity sales to take account of solar energy growth is to subtract from the econometric forecast of the levels of utilization of each fuel type on HVAC systems, based on historical relationships, a forecast obtained from a study of potential solar-market capture. The state-of-the-art techniques for conducting such studies are discussed later. At this point we need observe only that they will yield as their outputs forecasts of the number of homes with each alternative type of HVAC system, including solar, in every year from the present to the forecast horizon (H'_{it} for $i = 1, ..., m, t = 1, ..., h$), and that they require as inputs the projections of these same variables yielded by the econometrically estimated equations (H_{it} for $i = 1, ..., m, t = 1, ..., h$) and forecasts, based on historical relationships, of the number of newly constructed homes that would have installed a system of type i in year t (N_{it} for $i = 1, ..., m, t = 1, ..., h$).

The need to distinguish in the forecasts inputted into the market-capture study between the numbers of new homes and existing homes that will use each fuel type arises because the rate of introduction of solar space conditioning and water heating in new homes is likely to be considerably greater than in old homes. In new applications the full costs of solar will be compared with the full costs of the alternative systems, whereas in retrofit situations all or part of the existing conventional system's fixed costs are sunk and the full costs of solar will have to compete with the operating costs of the present system.

To forecast N_{it}, it will be necessary first to estimate a regression equation for the ratio S_{it}^N/S_{mt}^N which is defined analogously to S_{it}/S_{mt}, being the share of fuel type i relative to m of HVAC systems installed in the new homes in year t, rather than of the total sales of HVAC systems in year t. Specifically,

$$S_{it}^N = \frac{N_{i.}}{\Delta H_t + d_t H_{t-1}} \tag{3.7}$$

where, of course, $\Delta H_t + d_t H_{t-1}$ is equal to the total number of new homes constructed in year t which will install some type of HVAC system. Once market-share ratios S_{it}^N/S_{mt}^N for all $i = 1, \ldots, m - 1$ have been estimated, S_{it}^N can be calculated from the relationship

$$S_{it}^N = \frac{S_{it}^N}{S_{mt}^N} \frac{1}{1 + \sum_{j=1}^{m-1}(S_{it}^N/S_{mt}^N)}$$

and equation 3.7 can be used to estimate N_{it} for future periods.

If data on N_{it} for all $i = 1, \ldots, m$ are available for past years, an equation for S_{it}^N/S_{mt}^N can be estimated by regression the historical values of S_{it}^N/S_{mt}^N on the historical values of the same variables that appear in equation 3.2 for S_{it}/S_{mt}—omitting, however, the variables

$$R_{it} = \frac{r_{it}\,H_{i(t-1)}}{\sum_{j=1}^{m} r_{jt}\,H_{j(t-1)}}$$

and R_{mt} which measure the proportions of all present systems requiring replacement in year t that use fuel of types i and m, respectively.

In the absence of historical data on N_{it}, it will be necessary to estimate S_{it}^N/S_{mt}^N from the equation for S_{it}/S_{mt}. Any difference between these two ratios will be due largely to a difference between R_{it} and R_{mt}. If, for example, R_{it} greatly exceeds R_{mt}, then fuel type i accounts for a much higher proportion of systems requiring replacement in year t than does fuel type m, and its relative share of the total sales of HVAC systems is likely to be substantially greater than its relative share of space-conditioning and water-heating systems currently being installed in new homes. That is, $S_{it}/S_{mt} > S_{it}^N/S_{mt}^N$. Since a variation in the ratio R_{it}/R_{mt} from 1 is likely to account for the bulk of the difference between S_{it}^N/S_{mt}^N and S_{it}/S_{mt}, a reasonable estimate of S_{it}^N/S_{mt}^N could be obtained by setting $R_{it}/R_{mt} = 1$ in the regression equation for S_{it}/S_{mt} (equation 3.2).

Having obtained two estimates of the number of homes with each type of HVAC system in future years, one based on historical relationships and the other yielded by a market-penetration study, the forecaster can proceed in one of two ways to prepare this projection of total residential electric energy consumption. He can compute for each year t of the forecast period the sum

$$I_t = \sum_{i=1}^{m} C_{it}(H_{it} - H'_{it})_{it}$$

where C_{it} is an engineering projection of the electricity consumed by space-conditioning and water-heating systems using fuel type i in year t. Subtracting I_t from an aggregate forecast of residential electricity consumption produced by a regression equation estimated over the historical period would then yield a forecast adjusted for an expected acceleration in the rate of growth of solar space-conditioning and water-heating market penetration. To obtain accurate results, the HVAC system types $i = 1, ..., m$ must include all alternative types of solar systems, for example, those employing each of the different energy sources as a backup.

A key assumption underlying the method outlined above is that econometric forecasts of HVAC system stocks and of aggregate electricity consumption will be mutually compatible; that is, the sum $\sum_{i=1}^{m} C_{it} H_{it}$ is consistent with the forecast of total consumption. Both forecasts are assumed to fail to reflect the impact of solar-market penetration on sales to the same extent. An alternative approach which does not rely on this assumption but which, as previously mentioned, has problems of its own, is to build up a forecast of total residential consumption by aggregating the forecasts of the electricity consumed by the total stocks of each type of appliance including space-conditioning and water-heating systems. If this procedure were followed, the forecast of total consumption could be modified to take account of solar by simply adding $\sum_{i=1}^{m} C_{it} H'_{it}$ rather than $\sum_{i=1}^{m} C_{it} H_{it}$ to the electricity consumed by all other appliances.

Solar Energy and Peak Electrical Demand

Up to this point we have considered only the effect of the adoption of solar energy systems on kilowatt hour sales. However, as was observed at the beginning, utilities are most concerned about their forecast of peak demand because it determines their capacity requirements and their fixed costs. The problem of taking into account solar energy in a forecast of peak demand can be handled in much the same fashion as it was in forecasting kilowatt hour sales. A base forecast of peak demand which does not reflect the expected rapid increase in the adoption of solar energy must be made initially. Then the reduction in capacity requirements each year that will occur as the use of solar energy becomes more and more widespread must be subtracted from this base forecast.

Our discussion of the methodologies available for forecasting peak demand begins with those commonly used by the power companies and then considers various theoretical approaches discussed in the literature. It also includes some suggestions of our own.

In preparing hour-to-hour and day-to-day short-term power forecasts, most utilities add expected load changes to the current or recent-past loads. The changes in load are assumed to be functions primarily of weather conditions such as temperature, humidity, or wind velocity as well as historical hourly load patterns and day-of-week patterns.

Somewhat more sophisticated techniques are used to project loads over an intermediate period of 4 to 6 years and over the long run, that is, 10 to 30 years into the future. However, there is considerable variation in forecasting practice among the utilities. Many companies forecast annual system peak directly. But many others forecast components of the total load, such as the demand in different geographical areas or of different classes of customers, and then aggregate these demands, with adjustments for loss and diversity based on historical relationships. Another common practice is to forecast peak demand indirectly by projecting energy sales and the load factor.

The statistical methods employed in load forecasting generally involve some form of extrapolation. In many cases, least-squares regression is used to fit a time-series function which best represents the pattern of growth in the past to historical data. Some utilities determine the correlation between their loads and such fundamental determinants as income, industrial output, electricity rates, and so on. Most companies adjust the actual historic load data to reflect normal weather conditions. They project the peak demand which is most likely to occur if historical-average extreme weather conditions occur in the future.

An important class of approaches to peak-load forecasting that have received little practical application but have been discussed fairly extensively in the literature revolve on breaking down the load into base, seasonal or weather-sensitive, and in some cases irregular components. This method can be applied to forecasting demand in each hour of the day (and thus the entire 24-h load curve) or to predicting the peak demand during the day, week, or month.

It should be noted that seasonal variations in load are not solely functions of weather conditions. They are also due, in part, to the economic seasonal cycle and, of equal importance, at least for the winter months, to the change in the number of hours of daylight.

Weather sensitivity in the load is a result of an increase in demand for electric heating and power to operate furnaces and humidifiers, as winter temperatures drop below 60°F, and an increase in the demand for electricity to operate air conditioners, refrigerators, and freezers, as summer temperatures rise above 70°F. High humidity also increases the summer air-conditioning and refrigeration load while wind speed and cloud cover affect the demand for space heating and lighting during the winter.

The base component of the load is that portion of the total which responds solely to changes in such basic underlying economic determinants as incomes, industrial output, electricity prices, competing fuel prices, and the like and is insensitive to variations in weather conditions or other seasonal factors.

The division of the total load into its base and seasonal components can be accomplished in a variety of ways. One straightforward procedure, developed by Gupta (1969), requires seasonally adjusting the entire load by using a ratio-to-moving-average technique to obtain a seasonally adjusted base load which is then subtracted from the actual load to determine the seasonal component. The seasonally adjusted base load is forecast by fitting a trend curve to the historical observations. Weather models are estimated for the seasonal load by regressing seasonal demand on dry-bulb temperature measured as a deviation from its mean. This is done separately for the winter and summer months. The product of the weather variable and its coefficient in each weather function measures the weather-sensitive portion of the load, and the residual is equal to the non-weather-sensitive portion of the seasonal load. The summer and winter models are estimated by using weekly peak-demand data for a number of past years, with a separate equation being estimated for each year. The weather coefficient in these equations is then forecast into the future by time-series methods.

Another approach, also pioneered by Gupta, involves separating the total load directly into non-weather- and weather-sensitive components by simply regressing weekday peak demand on weather data to determine a winter and summer load model for each year. The intercept terms in the weather models represent the winter and summer base loads, and the products of the coefficient and the weather variables are equal to the winter and summer weather-sensitive components of the load. Since these models are estimated separately for each year, a value of the intercept term and of the coefficient of the weather variable is obtained for the winter and summer models for each year. These can be forecast into the future by using time-series techniques.

An alternative method for forecasting the weather- and non-weather-sensitive portions of the load that has not been discussed in the literature, but which appears to be an improvement over Gupta's technique, would be to correlate the base load (the intercept of the weather equations) and the coefficients of weather and seasonal variables with their basic economic and technological determinants.[1] The first step would be to divide the total load into its base weather-sensitive and non-weather seasonal-sensitive components by estimating the following equation for each year y of the historical period:

$$D_{yt} = a_y + b_{b_{1y}} T_w + b_{b_{2y}} W_{yt}^s T_s + b_{b_{3y}} S_{yt}, \qquad (3.8)$$

where D_{yt} = demand at the hour of peak demand in period t in year y

W_{yt}^w = value of winter-weather function at the hour of peak demand in period t in year y

T_w = 1 during the winter months; 0 during the summer

W_{yt}^s = value of the summer function at the hour of peak demand in period t and year y

T_s = 1 during the summer; 0 during the winter

S_{yt} = factor of seasonal variables including the number of hours of daylight

The weather variables W_{yt}^w and W_{yt}^s in equation 3.8 will be complex functions of temperature, humidity, wind speed, and cloud cover at and immediately prior to the hour of peak demand at time t. The functions for W_{yt}^w and W_{yt}^s are likely to differ substantially in form and in the choice of independent variables. Now W_{ty}^w may be equal simply to the difference between 60°F and ambient temperature; however, previous studies such as that of Corpening, Reppen, and Ringlee (1973) have shown that a much more complex weather function which includes humidity, cloud cover, wind speed, and so on, will produce superior results for the summer months. The vector of seasonal variables may contain, in addition to the hours of daylight, variables which measure the day-of-week effect on the load and the growth in the load over the year. Equation 3.8 can be estimated from observations of daily, weekly, or monthly peak demand in year y.

The base, weather-sensitive, and seasonal components of electric-utility loads have been changing in response to changes in such variables as incomes, population, appliance saturation levels, industrial output, rate of capacity utilization in industry, prices of competing energy sources, and electricity rates—especially demand charges. The final step in constructing the demand model is the estimation of equations explaining a_y, which represents the base load, and the weather seasonal coefficients b_{1y}, b_{2y}, and b_{3y} as functions of these economic variables. Since equation 3.8 will be estimated separately for each year of the historical period for which data are available, vectors of annual observations on a_y, $b_{1y'}$, $b_{2y'}$, and $b_{3y'}$, will be generated. These can be regressed on the annual values of the economic variables to estimate the explanatory equations that will be used to predict the values of the coefficients in future years. The probability distribution of peak demand in the future can be computed from equation 3.8 and the probability distributions of extreme values of W_{yt}^w and W_{yt}^s.

A different procedure from those discussed up to this point is presented in a paper by Chen and Winter (1966). In their model a forecast of peak demand, D_{t+1} on day t for day $t + 1$ is given by

$$D_{t+1} = B_{t+1} + W_{t+1} + T_{t+1}^* + B(CL_{t+1}) \qquad (3.9)$$

where B_{t+1} is the base load on day $t + 1$, W_{t+1} is the day-of-the-week effect on $t + 1$, and T_{t+1}^* is the difference between actual dry-bulb temperature and 65°F if the temperature is over 65°F or the difference between 50°F and actual dry-bulb temperature if the temperature is under 50°F. The cloud-cover variable CL_{t+1} is a dummy that takes the value of 0 for clear, 1 for partly cloudy, 2 for cloudy, and 3 for precipitation. Now B_t is projected in an adaptive forecasting framework as the weighted average of B_{t+1} and the difference between this period's actual load and the load predicted by all the other independent variables in equation 3.9; W_t is forecast in a similar manner.

Finally, mention should be made of a couple of methods suggested by Taylor (1975b). In the first he argues for the application of a Box-Jenkins time-series function to hourly load data in which the change in the load over a 24-h period is treated as a stochastic variable determined by an autoregressive moving-average process. The parameters of this process, Φ and K, can be estimated for a series of overlapping historical periods to generate a vector of time-series observations on Φ and K. His proposal is to employ a generalized least-squares procedure to estimate the functional relationships between Φ and K and their economic determinants: income, price elasticity, temperature, and so on. A generalized least-squares technique is necessitated by the autocorrelation among successive observations of Φ and K that arise because these variables were estimated for a series of overlapping periods. Taylor also proposes an econometric adaptive forecasting procedure similar to the adaptive technique used by Chen and Winters (1966), but with the base-load variable B_t being replaced by a vector of economic determinants.

Modifying the Base Forecast of Peak Demand to Take
Account of Solar-Market Penetration

Regardless of how the base projections of peak load are made, they must be modified to take into account the impact of solar energy utilization. From the market-penetrations study, estimates can be obtained of the number of homes that will employ solar energy for space conditioning and water heating in place of the alternative energy sources—principally electricity, natural gas, and heating oil. The difference between a solar home and a conventional one in terms of demand for power at system peak can be determined as was done in this book using a simulation model such as TRNSYS. Simulations should be performed by using weather data from the entire two- or three-week period during which the peak has occurred in substantial number of past years, so as to generate a probability distribution of demand at system peak of both the solar-heated and conventional homes. It can be shown that if the probability distributions of these two demands have the same variance and both are equally correlated with system demand, then the difference in demand is the difference between the means of the two probability distributions. This conclusion rests on the assumption that the utility regulates its capacity to keep the probability of loss-of-load constant at some well-defined critical level.

If the expected peak demand of HVAC system using energy type i is P_{it}, the total change in the expected peak-demand forecast for future year t by the base projections that will result from the growth of solar space conditioning and water heating, as predicted by the market-penetration study, will be

$$h = \sum_{i=1}^{m} (H_{it} - H'_{it}) P_{it}$$

As long as the variances of the probability distributions of demand are the same for solar and conventional homes, and both are identically correlated with system peak demand, the standard deviation of system peak will not be altered by the growth of SHACOB. Consequently, the utility will be able to change (hopefully reduce) its capacity by an amount equal to h without altering the probability of being able to meet peak demand.

To demonstrate this fact, let

K_t = total generating capacity in kilowatts in period t

O_t = forced outage in period t in kilowatts (generating capacity lost due to unexpected mechanical breakdown or other cause)

D_t system peak demand in period t in kilowatts

$A_t = K_t - D_t'$ the reserve-capacity margin available in period t to meet a forced outage

$f_O(O_t)$ = probability density function for O_t

$f_D(D_t)$ = probability density function for D_t in the absence of solar space conditioning and water heating

$f_D'(D_t)$ = probability density function for D_t given the expected market penetration of solar space conditioning and water heating

$\bar{D}_t$ = mean of expected value of D_t in the absence of SHACOB water heating, $\bar{D}_t = \int_{-\infty}^{\infty} D_t f_D(O_t)\, dD_t$

$\bar{D}_t'$ = mean or expected value of D_t given the expected market penetration of SHACOB

h $= \bar{D}_t' - \bar{D}_t$

σ_{D_t} = standard deviation of D_t without SHACOB

$\sigma_{D_t'}$ = standard deviation of D_t with SHACOB

ΔK_t = change in K_t made by the utility in response to expected growth in SHACOB

$P(A_t \geq O_t)$ = probability that A_t will exceed O_t in the absence of SHACOB

$p'(A_t \geq O_t)$ = probability that A_t will exceed O_t given solar SHACOB

In this analysis D_t and O_t are treated as stochastic variables. On the other hand, K_t is assumed to be determined solely by the utility. Our aim is to show that if there is no change in the standard deviation of peak load as a result of the growth SHACOB, then the probability of the utility reserve-capacity margin at time t being large enough to meet the forced outage occurring at time t will also be unchanged, provided that the reduction made by the utility in its generating capacity at time t is equal to the decrease in the expected peak demand due to the growth of SHACOB. In terms of the variables defined, if $\sigma_{D_t} = \sigma_{D_t'}$, then $P'(A_t \geq O_t) = P(A_t \geq O_t)$ provided that $\Delta K_t = h$.

It can be shown that

$$P(A_t \geq O_t) = \int_{-\infty}^{\infty} f_A(A_t)\, dA_t \int_0^{A_t} f_O(O_t)\, dO_t \qquad (3.10a)$$

and

$$P'(A_t \geq O_t) = \int_{-\infty}^{\infty} f_A'(A_t)\, dA_t \int_0^{A_t} f_O(O_t)\, dO_t \qquad (3.10b)$$

where $f_A(A_t)$ and $f_A(A_t)$ are the probability density functions for A_t with and without SHACOB, respectively. Since K_t is determined by the utility, it is not a stochastic variable and may be treated as a constant. Therefore,

$$\bar{A}_t = K_t - \bar{D}_t$$

$$\bar{A}'_t = K_t + \Delta K_t - \bar{D}_t \tag{3.11}$$

$$\sigma_{A_t} = \sigma_{D_t} \tag{3.12}$$

and

$$\sigma'_{A_t} = \sigma'_{D_t} \tag{3.13}$$

where $\bar{A}'_t$ and $\bar{A}_t$ are the expected values and σ'_{A_t} and σ_{A_t} are the standard deviations of A_t with and without SHACOB, respectively. Since $h = \bar{D}'_t - \bar{D}_t$, we know $\bar{D}'_t = \bar{D}_t + h$, and from equation 3.11 we have

$$\bar{A}'_t = K_t + \Delta K_t - \bar{D}_t - h$$

provided that $\Delta K = h$ and $\bar{A}'_t = \bar{A}_t$. Moreover, from equations 3.12 and 3.13 and the assumption that $\sigma'_{D_t} = \sigma_{D_t}$, we have $t'_{A_t} = \sigma_{A_t}$.

If $f'_A (A_t)$ and $f_A (A_t)$, the probability density functions of A_t, depend solely on their means and standard deviations, then $f'_A (A_t) = f_A (A_t)$ because $\bar{A}'_t = \bar{A}_t$ and $\sigma'_{A_t} = t_{A_t}$. Consequently,

$$P(A_t \geq O_t) = \int_{-\infty}^{\infty} f'_A (A_t) \, dA_t \int_0^{A_t} f_A (O_t) \, dO_t$$

$$= \int_{-\infty}^{\infty} f(A_t) \, dA_t \int_0^{A_t} f_A (O_t) \, dO_t = P'(A_t \geq O_t)$$

and the change in the utility's generating capacity which must be made in response to the growth of SHACOB in order to hold constant the probability of meeting system peak demand cannot be calculated in general without full knowledge of $f_D (D_t)$ and $f'_D (D_t)$. It was possible to do so in the case just examined because σ_{D_t} and σ'_{D_t} were assumed to be equal. However, there exists another special case in which the value of ΔK_t required to equate $P'(A_t \geq O_t)$ to $P(A_t \geq O_t)$ can be determined without knowing $f_D (D_t)$ and $f'_D (D_t)$; that is when σ'_{D_t} exceeds σ_{D_t} by enough to compensate for the decrease in the expected demand from $\bar{D}_t$ to D'_t, leaving $P'(A_t \geq O_t)$ when $\Delta K_t = 0$. The level of $P(A_t \geq O_t)$ considered acceptable by the utility will be very high, requiring that, since $A_t = K_t - D_t$, the probability be very small that $D_t \geq K_t - O_t$. The growth of SHACOB will permit no reduction in the utility's generating capacity [$P'(A_t \geq O_t) = P(A_t \geq O_t)$ when $\Delta K_t = 0$)] if the probabilities of values of D_t greater than $K_t - O_t$ are not reduced below the low level which the utility requires. This will occur if the demand for electric power of a solar home is likely to equal or exceed that of a conventional home under weather conditions that produce extreme values of D_t, system peak demand. On the other hand, if the expected difference between the demand of a solar home and of a conventional home is invariant with respect to whether the system peak is below, equal to, or above normal (that is, if the difference and the system peak are uncorrelated independent variables), then the growth of SHACOB will not alter the standard deviation of peak demand and, as has been shown, the utility will be able to reduce its capacity by an amount equal to the reduction in the expected peak due to solar energy utilization.

We are concerned, then, with two vital questions of fact. Are the standard deviations of demand at the time of system peak of solar and conventional homes the same? Is there zero correlation between the magnitude of system peak and the difference in demand at system peak between solar and conventional homes? Some light is shed on both questions by the results of the Feldman-Anderson study. The demands of solar and conventional homes at the time of system peak are presented for a half-dozen utilities in tables 11.4 and 11.7 and figures 11.14 to 11.18 of their report (Feldman and Anderson 1976a). It is evident that the standard deviations of demand of solar and conventional homes are approximately the same for the winter-peaking and one of the summer-peaking utilities (provided the very low demands of the solar home that occurred in some years are dropped from the sample). The result obtained with different collector areas indicate that this difference could be narrowed or even eliminated by use of a larger collector area and storage volume.

There was no direct information in the Feldman-Anderson report as to which years experienced the most extreme conditions in the sense of having the highest system peak demand relative to the normal peak demand. However, the year in which the demand of the conventional home was highest is indicated. Since there is probably a strong positive correlation between relatively high demand by conventional homes and relatively high system peaks, it is worthwhile to examine the relationship between maximum conventional home demand and the difference in demand between conventional and solar homes. The next chapter examines this relationship by using a case study that employs the systems model of chapter 2.

Note

1. To date, there have been relatively few econometric studies of peak demand. One of the few published articles on the subject is that of Cargill and Meyer (1971). They regressed deseasonalized monthly demand directly on a set of economic variables. Tolley, Upton, and Hastings (1977, pp. 29–37) used econometric and survey methods to estimate and project the contributions of individual household appliances to the system peak of the Wisconsin Electric Power Company.

4

The Simulation: A Case Study of Colorado Springs, Colorado

This chapter demonstrates the operation of the "systems" model in the case study of Colorado Springs, Colorado. In chapter 2 the model was described in detail. The model specifies the path to determine optimal investments in residential space-conditioning technologies, such as energy conservation and/or solar energy, from each of three perspectives, that of the homeowner, of the electric utility, and of society. The model identifies two major points onto which the optimization is focused. First, emphasis is placed on the building rather than partial components such as the solar energy or HVAC systems. This shifts the optimizing function away from only mechanical equipment to include contributions of building design and operation. Second, the model illustrates the important differences that exist among the homeowner, the electric utility, and society. It recognizes that homeowner or utility optimization does not necessarily coincide with the optimal configuration for society.

A Detailed Description of the Sectors for Energy Analysis

The general systems model presented in chapter 2 outlines the relationship among building, building owner, the electric utility, and the nation. More detailed discussion, however, is warranted. Each of the sectors delineated will be treated in detail. In addition, the operational model for the case study is selected and evaluated.

Case Study Sector

The utility selected is the City of Colorado Springs, Colorado. Weather data were obtained for Colorado Springs and Boulder, Colorado (the nearest weather site with yearly hour-by-hour radiation values). The Colorado Springs Department of Public Utilities is a municipal utility responsible for electricity as well as water, gas, and sewage.

The small size of the Colorado Springs service area makes the selection of weather location easy. Local climatological data for Colorado Springs do exist and contain all the required information, although cloud cover must be used instead of percentage of possible sunshine. The same is not true, however, for yearly weather sequences, where accurate solar radiation values are available nationally for only a few locations. The nearest available location for yearly weather including radiation is Boulder. Although Boulder and Colorado Springs are both at approximately 105°W longitude, there is a deviation between the weather at Boulder and Colorado Springs. The former is in the foothills while the latter is in the Front Range. Solar radiation is more available for Colorado Springs; however, Colorado Springs also has 800 (or 12 percent) more winter degree-days than Boulder (see ASHRAE 1977).

The important portions of the systems model have been identified as the annual building consumption in kilowatt hours, the building capacity demand in kilowatts, the cost-effectiveness to the consumer, the impact on utility load and costs, and the impact on society.

The Annual Consumption of Electricity

The annual performance of solar energy and energy-conservation investments can best be evaluated by reference to some standard for comparison. In this case, the base building (building 1) is taken to have a fully insulated 2 × 4 in wood stud wall, light thermal mass, 6-in of attic insulation, 16 percent double-glazed windows on the south wall, and no overhang shade. This building utilizes electricity for heating, cooling, and domestic hot water. The base building is a standard for all cost-effectiveness studies of energy-related factors in this book; thus it is important that its thermal performance be understood in some detail.

Values for the base building appear in table 4–1; configuration C, an all-electric version, is taken as a conventional building and serves as a basis for comparison. The table shows a total kilowatt hour consumption of 18,015 kWh, composed of 12,744 kWh for heating, 306 kWh for cooling, and 4,965 kWh for hot water.[1] Comparisons can be made to show the sensitivity of particular building design practices to annual electricity consumption. For convenience these alterations can be divided into active solar, building envelope, interior partitioning, and fenestration.

Active Systems

The all-electric base building 1C may be compared to other buildings and other HVAC system configurations. A combined solar heating and hot-water system with electric cooling and electric auxiliary, version A, replaces the all-electric system in version C. A solar hot-water system represents version B. The changes in the kilowatt hours of electricity are noted below. (See table 4–2.)

Table 4–1
Thermal Analysis

Utility:	Colorado Springs peak period: November to March, 4 to 9 p.m., weekdays
Building:	Base building 2 × 4, 6-in attic insulation, light room construction, no thermal shutters, without shade, 16% double glaze
Collector:	35.0 m², storage size 2.65 m³, 60° tilt

Building Demand at Time of Utility Peak; Building 1	
A: Solar space heating, solar hot water	4.6 kW
B: Electric space conditioning, solar hot water	5.5 kW
C: Electric space conditioning, electric hot water	5.8 kW
Base conventional building (used as comparison)	5.8 kW

Annual Building Electricity Consumption by Demand Period (kWh)			
	On Peak	Off Peak	Total
Building 1	(540 h)	(8,220 h)	(8,760 h)
A			
Electric for heat	164	1,801	1,965
Electric for cool	0	306	306
Electric for hot water	103	601	704
Solar pump electric	0	625	625
Total electric use	267	3,333	3,600
B			
Electric for heat	942	11,802	12,744
Electric for cool	0	306	306
Electric for hot water	0	2,887	2,887
Solar pump electric	0	65	65
Total electric use	942	15,060	16,002
C			
Electric for heat	942	11,802	12,744
Electric for cool	0	306	306
Electric for hot water	521	4,444	4,965
Total electric use	1,463	16,552	18,015

The solar heating and hot-water system reduces electricity consumption by about 14,500 kWh in the base building. This is more than an 8 percent reduction. The solar hot-water system, version B, reduces total kilowatt hours by 3,505, a reduction in electricity for hot water of 70 percent of the total space conditioning and hot water of 20 percent.

A further comparison of active systems is revealed by examining a range of collector and storage configurations. The building utilized is an energy-conservation building with fully insulated 2 × 6 in stud wall, 1-in Styrofoam sheathing, 12-in attic insulation, and an overhang shade. As revealed in table 4–2, the all-electric version utilizes 15,666 kWh annually while 35 m² of collector reduces this annual consumption to only 2,100 kWh.

Table 4–2
Effects of Active Solar Systems on Annual Kilowatt Hour Consumption for Space Conditioning and Hot Water
(*Energy-conservation building, Colorado Springs, with 16% glass, shade, 1-in total wall*)

Building [a]	Collector (m²)	Storage (m³)	Heating	Domestic Hot Water	Total [b]
4A	35	2.65	945	488	2,100
7A	35	7.95	392	527	1,650
8A	35	0.88	2,185	427	3,205
9A	17.5	1.32	3,815	1,296	5,501
10A	70.0	5.30	145	672	987
4B	5.6	Domestic hot water only	10,584	1,460	12,159
4C	None	None	10,584	4,965	15,664

[a] All building numbers refer to those descriptions in table 2–3.
[b] Includes electricity for solar pumps.

In this example, changes in the collector size do not radically change the performance of the active systems. Halving the collector size to 17.5 m² increases the electric consumption to 5,501 kWh, and doubling the collector only decreases consumption to 987 kWh. Much of the effect of larger collector sizes is negated by the electricity used for the pump which increases as the size increases. It is also important to note that storage/collector ratios have little effect on performance, as demonstrated by similar kilowatt hour consumptions in buildings 4A, 7A, and 8A.

The Building Envelope

Four insulating envelopes were modeled as shown in table 4–3. The most significant finding is the additional energy consumption of the minimally insulated building. The absence of any wall insulation (building 16) almost doubles the annual electricity consumption to 33,857 kWh from 18,596 kWh for the base building with shades. Inclusion of an additional 2 in of wall and 6 in of roof insulation in building 3 lowers consumption by 2,347 kWh to 16,290 kWh. An additional 1 inch of total wall insulation on the outside of the 2 × 6 in insulated stud wall (building 4C) reduces consumption by an additional 626 kWh.

The results vary slightly for the same set of buildings with 35-m² collectors. The larger building load of the minimally insulated building means a larger contribution for the active collection systems. Over 4,000 more kWh of useful energy resulted from a 35-m² collector on building 16 than on

Table 4–3
Effects of Insulation on Annual Kilowatt Hour Consumption for Space Conditioning and Hot Water; Colorado Springs, Colorado

Building	16% Glass, Shade	Heating	Cooling	Total[a]
2C	Base building	13,537	193	18,696
16C	Minimally insulated building	28,077	816	33,857
3C	Energy-conservation building without total wall	11,190	134	16,290
4C	Energy-conservation building with total wall	10,584	114	15,664

[a] Includes domestic hot water.

building 2. Active systems will appear more cost-effective when placed on the minimally insulated building. This is an example of the need to integrate all aspects of the building load. Cost-effectiveness of the entire building package (active system, energy conservation, and building design) is not guaranteed by the favorable performance of a single component.

Building Thermal Mass

Examinations of different levels of building thermal mass were made to determine the impact of thermal mass on annual consumption (table 4–4). Three levels of building thermal mass (low, medium, and high) were modeled. In otherwise identical buildings which had only 16 percent of the south wall in window and a shade, the heavy thermal mass of building 11 resulted in only very minor savings over the light thermal mass of building 4, conserving 300 kWh (2 percent). Although cooling consumption was cut by 69 percent, this represented an insignificant 67 kWh.

In two buildings with 80 percent south-facing window area and a shade but no thermal shutters, an increase from medium (building 5) to heavy (building 12) thermal mass produced somewhat improved results. Heating was cut by 1,271 kWh (14.8 percent) and cooling by 284 kWh (49 percent), and total consumption dropped by 1,555 kWh (10.6 percent).

Two buildings with 80 percent glass, shade, and shutters showed small savings with the increase from medium (building 6) to heavy (building 13) mass; 791 kWh (8 percent) was conserved, again with the cooling load cut considerably (47 percent), heating load cut relatively little (8 percent), and no change in hot-water demand.

Thermal mass affects building loads by dumping interior temperatures through its ability to store heat for later convective-radiation fluctuations. In the Colorado Springs heating season, for buildings with small glass area,

Table 4–4
**Effects of Thermal Mass on Annual Building Kilowatt Hour Consumption
for Space Conditioning and Hot Water; Colorado Springs, Colorado**

Building		Heating	Cooling	Total
4C	16% glass, shade (R24 wall, R38 attic) Light mass	10,584	114	15,664
11C	(R20 wall, R38 attic) Heavy mass	10,352	47	15,364
5C	80% glass, shade (R24 wall, R38 attic) Medium mass	9,179	581	14,725
12C	(R20 wall, R38 attic) Heavy mass	7,908	297	13,170
6C	80% glass, shade, winter shutters (R24 wall, R38 attic) Medium mass	4,654	869	10,489
13C	(R20 wall, R38 attic) Heavy mass	4,277	456	9,698

nearly all direct heat gain is immediately useful in meeting the heating load. Therefore, the storage capabilities of the thermal mass are not used often enough to produce a significant impact on the heating load. The capacity of the thermal mass is better used with 80 percent south-facing glass. This is because direct heat gain is much larger, resulting in more heat in the building than is necessary to meet immediate heating requirements. The presence of insulating thermal shutters appears to lower internal-temperature fluctuations and, in part, to duplicate the role of thermal mass.

Summer cooling loads, however, are significantly lowered by the introduction of additional thermal mass for all buildings studied. This is due to the ability of thermal mass to absorb excess heat for convection-reradiation during the cooler nighttime, thus keeping the interior within the comfort zone more constantly. In addition, the benefits of thermal mass at the time of utility peak demand, when temperature fluctuations are greater, will be demonstrated later.

Fenestration

The final area of examination is the use of windows and fenestration options to decrease heating and cooling loads (table 4–5). The fenestration designs that have been examined are overhang shading, window size, and

Table 4-5
Effects of Fenestration on Annual Kilowatt Hour Consumption for Space Conditioning and Hot Water; Colorado Springs

Building		Heating	Cooling	Total[a]
	Energy-conservation building with total wall, with 80% glass, shade, medium thermal mass			
5C	No shutters	9,179	581	14,725
6C	Winter shutters	4,654	869	10,489
	Masonry building with 80% glass			
12C	Shade, no shutters	7,908	297	13,170
13C	Shade, winter shutters	4,277	456	9,698
14C	No shade, winter/summer shutters	3,227	810	9,004
15C	Shade, no shutters, triple-glazed south windows	4,570	358	9,894
	Base building with 16% glass			
1C	No shade	12,744	306	18,015
2C	Shade	13,537	193	18,696

[a]Including hot water.

thermal shutters. Two buildings are modeled with and without an overhang shade: a conventional 2 × 4 in stud wall building and a passive masonry structure. In both cases, the shade functioned poorly. The base building (1) with a shade was compared to a similar building (2) with no shade. The shade did reduce 113 kWh from the cooling load. It also, however, added 793 kWh to the heating load, for a total addition of 681 kWh due to the shade.

The shade performed poorly in the masonry building as well. Cooling requirements were once again cut (354 kWh), but heating suffered badly, rising 1,049 kWh in building 13 (with shade) over building 14 (no shade). The total consumption increased by 674 kWh (7.5 percent). The results stem simply from the fact that the shade is sun-responsive rather than season-responsive. That is, if the shade blocks unneeded radiation in September, it will also block the radiation in the colder month of March when it is needed. This problem worsens in a passive solar building, particularly in Colorado Springs where direct solar heat gain comprises a greater proportion of the building load. An option that has not been examined but that may avoid some of these problems is a seasonally variant shade such as an awning or foliage-covered trellis.

Shutters were modeled on both well-insulated buildings and passive masonry buildings. In both cases, the shutters had a largely desirable effect. A building with a 2 × 6 in stud wall with 1 in of total wall, medium mass, and 80 percent south-facing glass was modeled without shutters (building 5) and with shutters that remained open at all times except wintertime with little or no insulation (building 6). The heating load decreased 4,525 kWh (49 percent), while the cooling increased 288 kWh (50 percent).

Dramatic load reductions were achieved when shutters were added to the masonry building. Building 12, with no shutters and 80 percent glass, consumed 7,908 kWh for heating; the addition of a winter shutter (building 13) decreased the heating consumption by 3,631 kWh (46 percent) and total consumption by 26 percent. A winter/summer shutter (building 14) with no shade decreased total consumption by 4,166 kWh. The inclusion of a summer shutter routine effectively eliminated all summer cooling loads. During the spring and fall, when temperatures fluctuate greatly, a significant amount of cooling was required in part because of a relative lack of sophistication in the controlling routine. A more sophisticated operational mode for the shutters could effectively eliminate this problem. It is unclear, however, whether this additional expense is a better system than human control. With a winter/summer shutter system such as was modeled, an air conditioner would be used during spring and fall, but never in summer. In reality, a resident would most likely open the window in lieu of running the air conditioner; such human intervention would effectively remove the need for better controllers.

The masonry building was also modeled with triple-glazed windows over 80 percent of the south wall (building 15). This reduced heating consumption by 3,338 kWh (42 percent) from that of the double-glazed building, but this load was higher than that for shuttered buildings. It is important to realize that triple glazing is always in place and requires no human or mechanical control. This convenience ensures its performance; with human or mechanical control unreliability or malfunction may cause the shutters to perform at less than optimal efficiency.

Insulating devices to reduce heat loss/gain through windows are very effective. Winter/summer shutters offer real promise because they provide shading as well; triple glazing may offer a reasonable compromise between the need for control and the desire for enabling direct heat gain while keeping heat loss down.

Two window areas on the south-facing wall were modeled, 16 percent and 80 percent of the wall. The smaller window area reduced cooling loads (building 11) from larger window areas (building 12), but the cooling load is a small part of the total load. In addition, this problem for 80 percent glass can be alleviated through the use of a summer shutter or seasonally variant shading such as a trellis or awning. These options, unlike fixed shading, also allow full winter use of the direct-gain capabilities of the large window area. The shutters are also capable of reducing the problem of heat loss through windows. Triple glazing offers a reasonable alternative, reducing both initial cost and required control, while enhancing direct heat gain.

One relevant point here is the confusing and often misleading use of percentages in exhibiting building performance. The solar energy contribution is often expressed as a solar fraction, a percentage of total energy use. This relative value does not express the actual number of kilowatt hours of electricity consumed and is therefore incapable of determining potential tradeoffs between various configurations. Solar fraction is particularly useless in translating a building's thermal performance measure into a cost-effectiveness measure. For instance, an 80 percent solar fraction for a 500-kWh load is insignificant in comparison to a 50 percent solar fraction on a 10,000–kWh load. Solar fraction also represents the average annual solar contribution and offers no suggestion as to the variance associated with this average annual value. Solar fraction and average annual consumption alone do not provide sufficient information concerning building loads to determine building impact on the servicing electric utility or the subsequent impact of load-variable tariffs on individual optimization strategy.

The Kilowatt Demand of Buildings at Utility Peak

A preliminary investigation by Feldman and Anderson (1976a) has indicated that active solar systems can impact negatively on utility loads and finances. The impact of passive solar and energy-conservation buildings must be ascertained. The critical variable in assessing the utility impact is the amount of electricity demanded by the buildings at the time the utility experiences system peak demand. At this point the need for more generation capacity, one of the largest utility expenses, is realized. The utility companies, through a number of new tariff designs, will pass these capacity costs back to consumers who are responsible for peak demands.

The peak demands in kilowatts for the representative buildings in Colorado Springs have been calculated in accord with the "long-run-costs" step in the methodology. The peak demands are displayed in table 4–6 and are differentiated by building type and configuration (A, B, or C).

For each building, the inclusion of a solar energy system reduces peak kilowatt demand. Type A, with the collector sized for heating and hot water, has a lower kilowatt demand than type B, with the collector sized for hot water only. Type C, an all-electric version, has the highest kilowatt demand of the three. Comparison of some of the different building types and solar equipment sizing provides an indication of capacity requirements. It is possible to isolate certain changes in building design such that all other design characteristics except one are held constant. In the following discus-

Table 4-6
Demands of Building in Kilowatts during Utility Peak; Colorado Springs

Building	Solar Heating and Hot Water (A)	Solar Hot Water and Electric Heating (B)	All-Electric Heating and Hot Water (C)
1	4.6	5.5	5.8
2	4.6	5.5	5.8
3	3.5	4.4	4.7
4	3.2	4.2	4.5
5	4.2	5.6	5.9
6	2.9	3.8	4.0
7	3.1	4.5	4.8
8	3.5	4.3	4.5
9	3.9	4.2	4.4
10	2.4	4.7	5.0
11	2.8	3.6	3.9
12	4.1	5.8	6.1
13	2.7	4.2	4.4
14	1.9	2.3	2.6
15	3.1	4.4	4.7
16	9.2	9.9	10.2

sion, the impact of changes in active systems, building envelopes, and fenestration are evaluated as to the impact on peak kilowatt demand.

Effect of Active Systems

In table 4-7 the kilowatt demand of an energy-conservation building with total wall and varying collector sizes is shown. A comparison of the building with no collector and 35 m² of collector reveals a one-third reduction in capacity requirement from 4.5 kW to 3.2 kW as a result of the active system. A problem arises in that the annual reduction in energy from this building is 90 percent, and thus the reduction in revenues to the utility under the present pricing tariff is significantly greater. The full extent of this phenomenon will be addressed in the sections on load factors and utility impacts.

These higher peak demands in relation to average demands are partly due to the weather conditions associated with peak conditions in Colorado Springs. The sky-cover factor during the peak days is less than 50 percent. Benefits from increases in collector size are constrained by lack of solar radiation during these critical times. This is evident from the performance of the buildings with other collector sizes. Doubling the size of the collector to 70 m² does not result in a 50 percent reduction in capacity requirement, but only a 25 percent reduction to 2.4 kW. A 17.5-m² collector increases the kilowatt demand to 3.9 kW.

Table 4–7
Variation in Collector Size on Kilowatt Demand; Colorado Springs

Building	Description	Demand[a] (kW)
	Energy conservation with total wall, 16 percent south glass, overhang shade	
10A	Collector = 70 m², storage = 5.3 m³	2.4
4A	Collector = 35 m², storage = 2.65 m²	3.2
9A	Collector = 17.5 m², storage = 1.32 m²	3.9
4B	Collector = 5.6 m² (hot water only)	4.2
4C	No active system	4.5

[a] Average sky cover is a measure of cloudiness and is used as a surrogate for percentage of possible sunshine when the latter is unavailable.

A possible solution to the problem of inadequate solar collection potential may be to increase the size of the storage. The sensitivity of variation in storage size on peak loads is not clearly demonstrated. Tripling of storage size from building 4 to building 7 has little effect on kilowatt demand. Simulation of buildings with storage sizes approaching seasonal storage has not been performed, but may result in reduced peak kilowatt demands.

Effect of Building Envelope

The impact of solar buildings on utility peak loads in Colorado Springs is a function of not only low levels of radiation, but also extreme ambient temperature. Changes in building envelope design may be more critical to the amount of capacity demanded than is solar equipment sizing. Building configurations incorporating energy conservation and passive design have been simulated to examine this hypothesis.

In table 4–8 a number of different building types have been modeled to determine the peak capacity requirements. The importance of insulation as a method of reducing kilowatt demands is quite evident. (The term *demand,* as used in this chapter, is an engineering term referring to kilowatts required; it should not be confused with its use in economics jargon.) An all-electric base building requires 5.8 kW of electricity capacity. By adding 6 in of insulation to the ceiling and utilizing a fully insulated 2 × 6 in stud wall, the capacity requirement is reduced by 1.1 kW to 4.7 kW. This finding is significant in that given the proper incentives, it could be less expensive for the utility to meet anticipated growth in peak demands by insulating homes in their service area than by constructing new generation facilities. This fact is evidence of the importance of conservation as a major contributor to the energy needs of this nation.

Another finding of note is the peak demand of the minimally insulated

Table 4–8
Variation in Insulation Levels on Kilowatt Demand; Colorado Springs

Building	Description	Demand[a] (kW)
Base building (2C)	R11, R19	5.8
Energy conservation without total wall (3C)	R19, R38	4.7
Energy conservation with total wall (4C)	R24, R38	4.5
Minimally insulated (16C)	R 4, R19	10.2
Masonry heavy-mass building (11C)	R20, R38	3.9

[a]All buildings are all-electric with 16 percent south glass with overhang shade.

building. The peak demand of 10.2 kW for the all-electric minimally insulated building with no wall insulation and a window shade is nearly twice the demand of the equivalent base building. The only change between the two buildings is the inclusion of 3.5 in of wall insulation in the latter building.

As expected by the diminishing-return nature of insulation, the increase in insulation beyond R19 walls and R38 ceiling does not result in as drastic a reduction of utility loads as does the initial increment. It is unclear whether the inclusion of an extra 1 in of total wall (Styrofoam) is justifiable. The determination of the optimal amount of insulation cannot be made by examining the peak capacity effects alone, but must include consideration of annual consumption and costs as well. This is discussed in the next section.

Another factor affecting the peak demand of a building is its thermal mass. A masonry structure retains more heat energy from solar radiation during periods of low temperature. The greater the mass, the greater is this potential. In table 4–9 the effects of thermal mass have been isolated. It can be seen that the higher-mass masonry building almost always has a lower demand than the better-insulated wood-frame building. In the first set of comparisons for the 16 percent south glass buildings, the heavy-mass masonry building 11 performs considerably better (3.9 kW to 4.5 kW) than the energy-conservation building without total wall (the 2 × 6 in wood-frame light-mass building 4). Even building 4, which is better insulated than the masonry building, has a higher peak demand. In the other cases, the differences between the wood frame with 80 percent glass and masonry demands are less. This is mostly due to the inclusion of additional thermal mass in the wood building.

It is clear from table 4–9 that the masonry building does reduce capacity requirements significantly. Building 11 performs better than the other buildings listed in table 4–8. The only difference between this building and the others is the inclusion of masonry walls and heavy building thermal mass. Building 11 performs so well that the capacity demand of the all-elec-

Table 4–9
Variations in Building Thermal Mass on Kilowatt Demand;
Colorado Springs

	All-Electric Configuration
16% glass, with shade	
Building 4 (R24, R38, light mass)	4.5
Building 11 (R20, R38, heavy mass)	3.9
80% glass, with shade, no shutters	
Building 5 (R24, R38, medium mass)	5.9
Building 12 (R20, R38, heavy mass)	6.1
80% glass, with shade, with shutters	
Building 6 (R24, R38, medium mass)	4.0
Building 13 (R20, R38, heavy mass)	4.4

tric version of building 11 is equivalent to that of building 9A (2 × 6 in stud wall with total wall R24 and a 17.5-m² collector). This is in spite of the fact that the costs for the masonry building are $1,800 less. The positive effects of heavy building mass are reflected in these results.

Effect of Fenestration

One of the particularly interesting results of the analysis was the impact of fenestration on capacity demands. The lowered capacity demands of many of the masonry buildings were reduced even further through utilization of south-facing glass and thermal shutters.

The first entry in table 4–10 shows the effect of adding thermal shutters (R11) to the energy-conservation building with total wall and 80 percent south-facing glass. The reduction in kilowatt demand is 1.3 kW for the active buildings and even higher for the all-electric version.

The benefits of shutters are more clearly shown for the masonry building (table 4–10). The additional investment in shutters lowered demand by 1.4 kW in the building with an active system and 1.7 kW in the building with all-electric heating and hot water. Triple glazing and a shade in the same masonry building with 80 percent south glass is almost as effective as the thermal shutters. It is concluded that use of large expanses of south-facing glass without the use of thermal shutters and/or triple glazing tends to significantly increase coincident peak demand. The benefits of large glass expanses with thermal shutters in reducing kilowatt demand are not ensured, however, since the kilowatt demands of buildings with these features are not

Table 4–10
Effect of Thermal Shutters on Kilowatt Demand; Colorado Springs

	$35\text{-}m^2$ Collector (A)	All-Electric (C)
Energy-conservation building with total wall, 80% south glass, with shade		
Building 5, without thermal shutter	4.2	5.9
Building 6, with thermal shutter	2.9	4.0
Masonry building, 80% south glass, with shade		
Building 12, without thermal shutter	4.1	6.1
Building 13, with thermal shutter	2.7	4.4
Building 15, with triple glazing	3.1	4.7

significantly less than those of similar buildings with only small (16 percent) glass expanses.

One passive technique, the overhanging shade, which is highly regarded by architects for its ability to reduce summer cooling loads, has a large negative impact on building demands during Colorado Springs utility peaks. The effects of the overhanging shade on kilowatt demands is shown in table 4–11. For the two examples of 16 percent south-facing glass, the shade has very little impact. The inclusion of an overhanging shade requires about 0.1 kW of electricity to compensate for a slight reduction in direct solar gain. The real effect of the overhang shade is realized when 80 percent south glass is utilized. In this case the overhanging shade eliminates the equivalent of 0.9 kW of direct solar heat gain. The 80 percent glass masonry building with winter/summer shutters and no overhanging shade (building 14) produces the lowest kilowatt demands of any building, 1.9 kW for the active system. The 2.6 kW demand for the all-electric version of this masonry building 14 is lower than any other building configuration except for energy-conservation building 2 with a 70-m² collector.

Financial Impact on Consumers

The thermal analysis provided an account of the flow of energy through the solar energy system and electricity through the building-utility system. At the same time, there is a flow of money and resources made in equipment investments and fuel to generate electricity. A financial analysis follows to determine the most appropriate combination of active solar, passive solar, and energy-conservation investments to minimize total costs.

In the case of residential energy use, several different perspectives can be identified because costs are not incurred directly by the end user. The cost of electricity for the user is not always recovered by utility tariffs which

Table 4–11
Effect of Overhanging Shades on Building Coincident Peak Demand;
Colorado Springs

	A (kW)	B (kW)	C (kW)
2 × 4 base building, 16% glass			
Building 1 (with shade)	4.64	5.69	5.82
Building 2 (without shade)	4.58	5.66	5.79
Minimally insulated building, 16% glass			
Building 16 (with shade)	9.16	10.10	10.22
Building 17 (without shade)	9.07	10.02	10.15
Masonry building, 80% glass			
Building 13 (with shade)	3.74	4.31	4.43
Building 14 (without shade)	1.86	2.46	2.58

apply to broad ranges of users. There are costs, as well as some benefits, which are not, however, reflected in electricity prices. These include dependence on foreign fuel supplies, the social cost of embargoes, and environmental degradation. The price a consumer pays as a determinant in his decision making does not always lead to minimum total costs for society, as we examine below.

A procedure for accounting for the cost of each building from various perspectives has been developed. Table 4–12 details the incremental costs of modifications to the base building (a 2 × 4 in insulated wood frame, with 6–in attic insulation). The base building serves as the basis of comparison with all other buildings. Three different supply configurations have been examined for the base building. They include (A) a solar combined hot-water and heating system with solar collector sized for both heating and hot water; (B) a solar hot-water system with electric heat, with the solar collector sized solely for hot water; and (C) a conventional electric hot-water and heating system. All utilize a compression air conditioner. The incremental cost of each building is listed in the table. These costs are derived from Means (1976) and represent the base assumption. Sensitivity analysis is performed below on these costs.

First-Year Cost Analysis

The analysis presented in table 4–13 is for first-year cost analysis.[2] The same assumptions have been used to examine the financial situation of these buildings under life-cycle cost analysis. Both modes of analysis are relevant since there is no consensus as to which better reflects the investment deci-

Table 4–12
Building Costs; Colorado Springs
(*1975 dollars*)

	Incremental Cost over Base Building	
	16% Glass	*80% Glass*
Wall Models		
2 × 4	− 407	
2 × 4, insulated	0	
2 × 6, insulated	256	932
2 × 6, insulated, 1–in total wall	663	1,182
8–in masonry, 4–in total wall	3,183	2,728
Roof Models		
Standard roof	0	
Asymmetrical roof supporting 17.5-m² collector	878	
Asymmetrical roof supporting 35-m² collector	797	
6–in additional ceiling insulation	464	
Solar System Models		
Collectors	160/m²	
Fixed collector costs	1,687	
Storage size		
Domestic hot-water system	2,500	
Fenestration Models		
4–in overhang shade	1,108	
Shuttering (with 80% south wall windows)	1,015	
Roller blind shades (80% windows)	120	
Medium mass	700	
Triple glazing (80% south wall windows)	1,007	
Load Management	2,500	

sions of homeowners. Life-cycle costs do allow the consumer to anticipate a rise in the cost of fuel and to plan accordingly. For a number of reasons, however, the individual probably does not base his investment decisions on a 20–year horizon. (One other method of accounting for the consumer's lack of concern with the extended future is the use of very high discount rates.)

The analysis begins with an assessment of the costs to the consumer in the first year. The sum of the amortized yearly investment in capital equipment, the electric bill, and maintenance costs is the total cost incurred by the consumer. Incremental capital costs are calculated as those investments beyond the standard equipment in the all-electric base building. Should any of these buildings with additional investment for energy reduction have a

Table 4–13
Financial Analysis: First-Year Costs
(*1975 dollars*)

Utility:	Colorado Springs, peak period: November to March, 4 to 9 p.m., weekdays
Building:	Base building 2 × 4, 6–in attic insulation, light room construction, no thermal shutters, without shade, 16% double glazing
Collector:	35.0 m² storage size 2.65 m³, 60° tilt

Type		Incremental Building Cost(s)
A:	Solar space heating, solar hot water	8,254
B:	Electric space conditioning, solar hot water	2,500
C:	Electric space conditioning, electric hot water	1
D:	Base conventional building (used for comparison)	0

Analytical Assumptions

Life Cycle: 20 yr	Mortgage: 20 yr
Discount rate: 0.08	Rate: 0.085
Electricity price escalator: 0.070/yr	Downpayment: 0.00
Fuel cost escalator: 0.100/yr	Salvage value: 0.00
Cross-elasticity between peak and off-peak periods: 0.0	

Building	A		B		C	
Rate Schedule	*AC*[a]	*MC*[b]	*AC*	*MC*	*AC*	*MC*
Incremental						
Capital cost	872	872	264	264		
Electric bill	90	82	400	337	450	420
Maintenance	83	83	25	25	0	0
Total cost (consumer)	1,045	1,037	689	626	450	420

[a] AC = existing rate schedule

[b] MC = simulated marginal-cost rate schedule using the Cicchetti, Gillen, Smolensky model.

lower first-year cost than the base building, then this would signal a very favorable investment. This would mean that even in the initial year of a long-term investment, the entire yearly costs of the investment, including the interest, would be less than the savings in electricity costs.

The total cost to the consumer for base building A under the existing rate schedule is $1,045 (table 4–13). This figure represents the first-year expenditure for the incremental amortized capital costs of a solar combined hot-water and heating system ($872), electricity costs for electric air conditioning and auxiliary electricity for the solar systems ($90), and maintenance ($83).

The cost-effectiveness of A (a combined solar hot-water and heating system on the base building) must be evaluated in comparison to configuration C (a conventional electric and hot-water system). In this case the cost to the consumer for A ($1,045) is greater than the cost for C ($450), and this signals that an investment in a combined solar hot-water and heating system is not desirable by the building owner under this particular set of assumptions. The first-year cost to the consumer for the hot-water system on the base building (B) is also higher ($689).

While the solar energy system on the base building is not cost-effective, on other buildings it may be. A summary of the costs for each of the buildings is presented in table 4-14. This table examines the costs under existing average cost rates. The cost to the consumer in column one shows that the $450 for base building 1C is the lowest value. The next two closest buildings are also all-electric versions. A base building with an overhanging shade (2C) increases the first-year cost by $74 to $524, and thus the inclusion of the shade on the base building is not an effective investment. Another building, the energy-conservation building without total wall (3C) had a first-year cost of $542. To evaluate the cost-effectiveness of the additional insulation, the effect of an overhanging shade must be controlled; therefore comparison must be between the two buildings with shades. The incremental investment in insulation beyond the base building 3C versus building 2C results in only a $18 increase in total costs in the first year. With any increase at all in the price of fuel, this would become a feasible investment.

Table 4-14
Summary of Financial Analysis—No Tax Credits—First-Year Costs
(*1975 dollars*)

Building	Solar Heating and Hot Water (A)	Solar Hot Water Electric Heating (B)	All-Electric Heating and Hot Water (C)
1	1,045	689	450
2	1,106	729	524
3	1,150	745	542
4	1,186	772	569
5	1,360	717	534
6	1,425	889	686
7	1,338	773	570
8	1,159	754	551
9	918	760	557
10	1,929	773	570
11	1,447	1,031	828
12	1,405	928	725
13	1,502	959	756
14	1,447	834	680
15	1,443	856	702
16	1,362	1,067	864

Additional investment in insulation (a 1–in sheathing of Styrofoam) in the energy-conservation building (4C) results in costs that are slightly higher than for the energy-conservation building without total wall (3C). Investments in the all-electric masonry buildings (11C, 12C, 13C, 14C, and 15C) are considerably higher. The first-year costs for the active configurations (A) are even higher still, generally twice the cost of an identical building with all-electric space heating and hot water.

Life-Cycle Analysis

The fact that the first-year cost analysis yields higher costs for solar and energy-conservation investment does not bode badly for consumer investment in these alternatives. Annual increases in all fuel costs which are expected in the future will make some investments cost-effective over their life cycle. In life-cycle cost analysis, the method utilizes as a basis those assumptions adopted by the Department of Energy (Bezdek 1978). These assumptions are:

Life cycle = 20 yr

Discount rate = 8 percent

Interest rate = 8.5 percent

Downpayment on incremental construction costs = 20 percent

Rate of annual increase in fuel cost = 10 percent

Rate of annual increase in electricity price = 7 percent

Maintenance costs are assumed to be 1 percent of capital costs of active systems and thermal shutters and to increase at the inflation rate (6 percent per year). We assume that maintenance costs for insulation, windows, and building structure remain constant for each building type. No salvage value for the solar energy system after the 20–yr lifetime is assumed. An actual salvage value greater than zero would tend to make solar investment more cost-effective.

One of the problems with utilizing life-cycle cost is the need to provide information concerning future costs. Different perceptions exist with regard to future prices and the benefits accrued from future streams of income in comparison to present streams of income. Thus it is difficult to specify the appropriate price escalation rate and discount rate. Three future scenarios have been identified in which the benefits of the future and energy conservation assume increasing importance. As a result, the labels *low, medium,* and *high* have been attached to the scenarios:

1. Low: discount rate (DR) = 10 percent, electricity increase (EI) = 6 percent
2. Medium: DR = 8 percent, EI = 7 percent
3. High: Dr = 6 percent, EI = 10 percent

Low Scenario. As was the case in the first-year cost analysis, the all-electric base building (1C), without the shade, has the lowest life-cycle cost under the low scenario. The energy-conservation building 3C is now the second-lowest cost alternative and lower cost than the base building with a shade. This implies that the additional insulation is a justifiable investment, while the investment in the shade is clearly not justifiable. The other energy-conservation buildings, 4C and 6C, are similarly cost-effective as are two masonry buildings 14C and 15C. Active solar investment does not appear cost-effective under this scenario.

Medium Scenario. Although the all-electric base building 1C retains its position as the lowest cost alternative, the gap is less between that building and others. More importantly, several buildings with shades are less expensive than the base-building version with shade. This comparison illustrates that under this scenario both 2 × 6 in walls and 12-in attic insulation and 2 × 6 in walls with total wall and 12-in attic insulation are cost-effective over standard insulation.

This scenario is equivalent to the one used by Bezdek (1978) which found solar hot-water heaters to be cost-effective in comparison to electric hot-water heaters in Grand Junction, Colorado. The low cost of electricity for Colorado Springs—$0.025 per kWh as compared to $0.04 per kWh in the Bezdek study—does not support this finding in Colorado Springs. In fact, the cost of solar systems was significantly higher than any of the all-electric energy-conservation and masonry buildings.

High Scenario. This scenario gives the most value to the future of any of those examined, and as a result, findings are significantly different under this scenario. The all-electric masonry building 14C with differentiating winter/summer shutters on 80 percent south side is the lowest cost alternative. Following closely is another masonry building with a permanent shade and 80 percent south triple glazing. Two other buildings which utilize passive design (80 percent glass and shutters), 13C and 6C, also had lower costs than the base building 1C. A large number of buildings—twelve—had lower costs than the all-electric base building with shade, including two with solar hot-water systems (14B and 15B) and one with solar combined system, a 17.5-m² collector on an energy-conservation building (9A). One finding is that it is better to place a 17.5-m² collector on an extremely well-insulated building than it is to place a 35-m² collector on the base building. Unfortunately, it is not clear from the analysis what the cost-effectiveness of a 17.5-m² collector is on the base building.

A finding worthy of note is the cost analysis for the minimally insulated buildings. The simple investment in wall insulation cuts the life-cycle cost almost in half. There appears to be no clearer savings derived from any other investment. Contrarily, an extremely large investment such as a 70-m² collector system and additional insulation as in the energy conservation 10A does not guarantee a savings, for this building is one of the highest in cost. In addition, as would be expected, investment in an active system on masonry buildings is not justified.

Impacts on the Utility

These cost-effectiveness evaluations provide an indication of benefits accrued to individual consumers as a result of investment in solar energy and/or energy conservation. Other impacts not accounted for in the consumer analysis also exist, including impacts on electric utilities that provide backup power or that compete for energy sales. The true societal value of a new energy source must include these costs or benefits in the final cost analysis.

The introduction of new building designs or technologies will affect a utility by altering the utility load characteristics. It is important for the utility to be able to assess the extent of this impact in order to adjust long-range capital expansion plans. In addition, a measure of the cost to service these new technologies must be obtained in order to evaluate the appropriateness of particular rate schedules.

Load Factors. The thermal analysis so far calculated consists of two separate pieces of information: the kilowatt demand at the time of the utility peak and the annual consumption in kilowatt hours. Individually, neither of these values provides information concerning building thermal performance that is adequate for calculating building cost-effectiveness or building impact on utility. Combining the two values into one value called *load factor* does aid in comprehending building thermal performance, particularly with regard to utility impact.

The individual building load factor—the average kilowatt demand for space conditioning and hot water divided by the kilowatt demand for space conditioning and hot water at the time of utility peak load demand—is an indication of the impact of a particular building on the utility load. This value is useful in providing a simplified indication of utility generation efficiency, but it should be noted that it does not provide an analysis of utility finances. (Because load factor is merely a ratio and not an absolute value, it cannot predict the absolute level of impact.)

The load factor for the Colorado Springs utility as a whole was 63 percent for 1975. The all-electric base building had a load factor of only 36 percent. This indicates that this aspect of building load would worsen the overall load factor and operating efficiency of the utility. Higher load factors would translate into more constant use of generation facilities and therefore more efficient system operation.

The load factor for each building modeled is shown in table 4–15. Clearly the most significant finding is the extremely low load factors for the buildings with active systems. Most buildings with solar space- and water-heating systems had load factors lower than 10 percent. This strikingly illustrates the cause for concern by utilities over the massive introduction of solar energy systems with electric backup. It appears that in Colorado Springs a large amount of the auxiliary energy required by the buildings with active systems is actually demanded at times of utility peaks. The larger the active system, the truer this is.

Several other significant trends are noticeable concerning load factors. All but one of those buildings that utilize large amounts of passive direct solar heat gain also have lower load factors. It is logical to assume that cloudy conditions that exist at times of utility peak will affect both active and passive solar collection. Counterbalancing the trend of low load factors

Table 4–15
Individual Load Factors; Colorado Springs

Building	Solar Heating and Hot Water (A)	Solar Hot Water Electric Heating (B)	All-Electric Heating and Hot Water (C)
1	0.09	0.33	0.36
2	0.09	0.31	0.37
3	0.08	0.33	0.40
4	0.08	0.33	0.39
5	0.11	0.11	0.18
6	0.07	0.21	0.30
7	0.06	0.31	0.37
8	0.11	0.31	0.38
9	0.16	0.32	0.39
10	0.05	0.30	0.36
11	0.08	0.38	0.45
12	0.06	0.19	0.25
13	0.05	0.17	0.25
14	0.09	0.29	0.40
15	0.05	0.17	0.24
16	0.19	0.35	0.38

for solar buildings are the improved load factors of energy-conservation buildings. The use of attic and wall insulation beyond the base building improves the load factors of the all-electric buildings. One unique building is the masonry building with 80 percent south glass, a winter/summer operating shutter, and no shade. The load factor of this building is larger than that of the base building despite the dependence on passive heat gain for a large portion of building heating load. The inclusion of effective shutters (which operate on a seasonally differentiated basis) and heavy thermal mass provides an energy-conserving device which counteracts any bad effect noticeable in other passive buildings.

The true extent of the impact of these buildings on the utility company cannot be assessed merely by the use of the load factor. An example of this is the similar load factors obtained by buildings 4 and 16. Building 4 has a yearly energy consumption of less than one-half of building 16 and thus has a completely different absolute effect on the utility loads and finances. This is shown more clearly below.

Impact on Utility Cost/Revenue Balance. The utility's cost of service is affected by drops in demand and consumption. The actual dollar impact of these changes is not often easily obtained nor comprehended, because of the complexity of finances within the utility's position as a regulated industry. The financial impact on the shareholders of a private utility is a separate issue from impacts on the customers and the rates they are charged.

In general, the impact of individual loads on the overall revenue situation of utilities is not critical. Utility profits are determined through the regulatory mechanism where total system operation is concerned. Rate schedules are created not to produce equivalent profits from each kilowatt hour sold, but rather to combine broad ranges of users into rate classes. Some discrimination is inherent in any tariff and, in fact, is often justified by the regulatory commissions for other reasons of public good. The impact of a particular part of a customer class load must be evaluated. If the cost to service that load is greater than the revenues received via the electric rate, then those particular users are being subsidized by the rest of the utility customers. Examination of the impact of the various building configurations demonstrate this more clearly.

The total cost of service to the Colorado Springs utility for the all-electric base building is $396 for the first year ($199 in variable costs and $197 in capacity costs). The revenue received by the utility under the average cost-declining block tariff is $450. This leaves a balance to the utility of -54, meaning that the cost of service for the all-electric base building is exceeded by the electric bill by $54. In table 4-16 the first-year cost balance to the utility is listed for each building from the existing average cost-declining tariff block.

Table 4-16
First-Year Cost Balance to Utility
(*1975 dollars*)

Building	Solar Heating and Hot Water	Solar Hot Water Electric Heating	All-Electric Heating and Hot Water
1	106	−37	− 54
2	107	−22	− 62
3	85	−30	− 68
4	78	−27	− 64
5	87	112	74
6	72	29	−9
7	83	−17	− 55
8	73	−16	− 54
9	65	−23	− 60
10	67	−11	− 48
11	68	−44	− 82
12	110	62	24
13	75	53	16
14	44	−4	− 38
15	88	60	22
16	94	−87	−124

Two major trends in the first-year cost balances can be identified. The most important is the large negative values for most of the all-electric buildings (C) and some of the buildings with solar hot-water heaters (B). Building configurations with high electric consumption and high load factors have negative cost balances. A negative value implies that the electric rate recovers, via the electric bill to the consumer, more revenue than is estimated for the utility's cost of service.

In contrast, high positive cost balances for configuration A buildings (the solar heating and hot water) are observed. These results illustrate the revenue problem envisioned by utilities that must supply auxiliary power to solar systems. For example, the estimated cost of service for the active solar version of the base building in the first year is $196 ($40 for variable run and $156 for capacity costs), but the revenue is only $90. In this example, the all-electric buildings are subsidizing the solar buildings. In comparing the impact on the utility of the all-electric and solar base buildings, the difference in net revenue to the utility is $160.

Societal Impact

The need for generation capabilities at system peak has been cited as a major problem for electric utilities. The opportunity for solar energy and/or energy conservation to benefit the utilities and society stems from the ability of these technologies to provide generation capabilities at a lower cost or curtail the need for the facility.

Certain barriers exist which prevent utility companies from fully exploring and utilizing these decentralized investments, despite clear financial benefits. Regulated utilities are not managed by the principle of cost minimization in which the less expensive alternatives are considered superior investments. In spite of market failures for individuals, either consumers or utilities, societal welfare is best served via minimization of the total costs.

At present the cost of an incremental kilowatt of capacity for Colorado Springs is $34 per year (1975 dollars). Compared to the typical U.S. utility, the cost of generation for Colorado Springs is low. In table 4–17 the costs of several alternative technologies are compared to the $34 per kilowatt per year for conventional sources. The table illustrates the clear savings that the insulation of minimally insulated buildings has over new generation capacity. The $41 per year for insulation saves the utility $150 in generation costs alone. Savings in fuel costs are additional. The $41 represents the cost of wall insulation in new construction. For retrofit purposes this cost would be considerably higher but would in all probability still represent a direct savings in capital generation costs.

Other investments in insulation and solar energy do not produce one-to-one cost savings in generation, and therefore investments in these alternatives cannot be justified solely on the basis of forestalled generation capacity. This, however, does not preclude the feasibility of these alternatives based on a combined reduction in capacity and fuel. In addition, the differences between the costs of capacity are not so great that changes in cost of conventional electric generation capacity might not make these alternative investments viable. For example, had the $34 per kilowatt per year value for Colorado Springs been above $69 (a value well below that of many utilities), then utility investment in insulation up to 2 × 6–in stud walls and 12–in attic insulation could be justified on savings in capacity costs alone. It does not appear, however, that in Colorado Springs investment in active solar energy by the utility can be translated into a significant reduction in conventional generation capacity.

The other aspect of utility costs is the variable (energy) costs. In Colorado Springs these costs have been determined as $0.013 on peak and $0.009 off-peak. These values for variable costs are also very low in comparison to other utilities. This makes Colorado Springs a reasonably good example to use since its marginal costs per British thermal unit are not that much higher than the costs of natural gas in that part of the country. One can then begin to think about cost comparisons with gas utilities. While gas utilities do not have the same magnitude of peaking problems, the approach is not too dissimilar to that used here.[3]

Table 4-17
Comparison of Costs per Kilowatt between Conventional Generation Technology (Coal) and Decentralized Alternatives; Colorado Springs
(*1975 dollars, new construction*)

Investment	Kilowatts Saved at Residence over Base Building	Savings of Utility Generation Capacity (Kilowatt Saved × $34/(kW · yr)	Savings of Variable Costs to Utility	Total Savings in Cost to Utility	Annual Amortized Cost to Consumer of Decentralized Capacity (New Construction)
Insulation of uninsulated 2 × 4 wall (buildings 16, 2)	4.4	150	162	312	51
Insulation to 2 × 6 6-in additional attic insulation	1.1	37	27	64	76
Masonry, 80% glass and thermal shutters (buildings 1, 14)	3.2	109	99	208	445
Solar hot-water system versus conventional domestic hot water	0.3	310	39	49	264
35-m² collector, 2.65-m³ storage on base building	1.2	41	175	216	873

Including savings in variable costs to the utility along with savings in kilowatt capacity costs provides a more complete indication of benefits of decentralized investment. In table 4-17 the fuel savings that accrue from investment in new construction of 3.5 in of wall insulation ($162) is commensurate with cost savings in capacity ($150). Together a savings in utility costs of $312 per year is realized from an amortized investment of only $51. For retrofit purposes it appears as though costs of retrofitted wall insulation can be eight times higher ($320 per year, or $3,000 total investment) and still produce a savings.

Other investments in Colorado Springs do not result in such significant benefits. Large amounts of insulation beyond the base building are not justified solely on utility savings in the first year. It is probable that given the rise in fuel costs and inflation, this particular investment will produce greater dollar savings to the utility than an investment in conventional generation facilities.

Solar energy investment in either a hot-water system or a 35-m² collector system does not produce an equivalent ratio of investment to benefits as results from those investments in energy conservation. A fuel savings of $175 results from an investment in the 35-m² combined hot-water and space-heating system. However, a lack of corresponding reduction in incurred capacity produced only a $41 savings in generation capacity costs for a total savings of $216, despite a yearly investment of $873.

This analysis provides only a partial picture of societal benefits. It is inappropriate to evaluate societal benefits over a one-year lifetime, particularly when it is desired that the nation's well-being exist even longer than any particular investment lifetime. Also, not all costs—such as the impact on foreign dependence on energy supply, balance of trade, and the environment—have been accounted for. These social costs are difficult to quantify and are often ignored completely; yet their cost is significant. [For an example, see Costello (1976, p. 22) where a range of $55 to $330 in benefits to the nation per kilowatt of capacity forestalled was estimated.] The real value of solar energy and energy conservation remains a political issue. Table 4–18 shows the societal benefits of different investments from a 20–yr life-cycle perspective. Values exhibited here represent a very short-term perspective as to societal benefits. Funds are revealed which provide the direction as to maximum benefits. External benefits are not quantified, but the savings in kilowatts and kilowatt hours is noted.

Clearly, the benefits to society of investments in wall insulation in poorly insulated homes are shown. Societal savings of $6,600 and 4.4–kW capacity and 300,000 kWh in fuel are realized for an initial investment of $390. The cost of retrofiting existing underinsulated housing stock, while considerably higher, would still appear to provide societal benefits far exceeding costs. Both the inclusion of additional insulation and passive solar also obtain benefit/cost ratios exceeding 1.

A real key to societal benefits is the measurement of fuel saved. Investment in wall insulation produces about the same fuel savings as investment in a 35-m² collector system despite a cost ratio of over 20:1. The reduction in capacity (kilowatt) requirement as a result of the investment in insulation (4.4 kW) is almost four times greater than the reduction in kilowatts from the active solar investment (1.2 kW).

Societal savings in fuel from additional energy insulation and from investment in a solar hot-water system are equivalent. In addition, the investment in energy conservation reduces kilowatt capacity requirements of the utility 1.1 kW for the insulation to only 0.3 kW for the solar hot-water system.

Table 4–18
Comparison of Societal Benefits and Costs among Various Technological Investments; Colorado Springs[a]

Investment	Total Life-Cycle Cost to Consumer per Investment	Total Life-Cycle Savings to Utility in Dollars per Investment	Ratio of Savings to Costs	Kilowatt Hours Saved between Alternatives per $10,000 Invested	Kilowatts Saved between Alternatives per $10,000 Invested
Insulation of uninsulated 2 × 4 wall (building 16 versus building 2)	500	6,600	14:1	7,769,000	110.0
Insulation to 2 × 6 and 6-in additional attic insulation (building 2 versus building 3)	900	1,300	1.5:1	667,000	15.0
Masonry, 80% glass and thermal shutters (building 2 versus building 14)	4,500	4,500	1.0:1	531,000	9.0
Solar hot-water system versus conventional domestic hot water	3.100	800	0.3:1	160,000	1.0
35-m² collector, 2.65-m³ storage on base building versus conventional electric building	10,100	5,100	0.5:1	349,000	1.5

[a] 1975 dollars, social discount rate = 6%, rate of fuel escalation = 10%, incremental cost of generation capacity = $34 per year; assumes new housing costs only.

Summary of the Case Study

The Colorado Springs case study demonstrates the importance of the systems model in determining optimal building space conditioning. The model identifies the tradeoffs among the many building options, including energy conservation, solar energy, and building design. The model also differentiates among the perspectives of the homeowner, the electric utility, and society. Major findings of the case study are summarized here.

The model emphasizes the importance of the building as the central focus of the system. It is the building load, and not a subset of that load (such as the solar system), that is important. An example of this is the relationship between active solar performance and building loads. In a minimally insulated building, a 35-m² collector/storage system provided over 4,000 kWh more useful energy than it did in the base building. The economics of that solar system would appear more favorable in the minimally insulated building. In considering the entire load, however, the investment in insulation appears to be far more favorable.

The actual cost-effectiveness for the homeowner depends on the financial parameters which best reflect his personal future investment outlook. The standard 2 × 4 in fully insulated stud wall appears to have the lowest cost in the first year. Rising fuel costs, however, provide future benefits to the consumer for energy savings investments. By using several scenarios of the benefits of future costs and savings, the life-cycle cost to the consumer has been calculated. The results indicate that energy conservation to at least a 2 × 6 in fully insulated stud wall and 12–in attic insulation becomes the preferred investment under the least future-oriented of the life-cycle scenarios. The most future-oriented scenario examined favors combined investments in masonry walls, large south-facing window area, and mechanical shutters.

Investment in active solar energy systems does not appear to be justified even under the most favorable economic assumptions. Similarly, the absence of wall insulation appears to be a significantly costly practice. For example, in the high future scenario where the discount rate is 6 percent and the rate of electricity price increase is 10 percent, the life-cycle cost of the minimally insulated building is twice the cost of a similar building with wall insulation (the base building). It also does not appear as though overhanging shades are a particularly good investment.

The average annual load has been shown to provide insufficient data on building performance. There is a need to determine building performance at time of utility peak loads so as to assess the building impact on the utility and the subsequent impact of new utility rates on building cost-effectiveness.

A significant finding of this book is the variation in performance of buildings at utility peak. Solar energy systems do have lower peak demands than similar all-electric buildings. The kilowatt demand of the 35-m² collection system on the base building lowered peak demand from 5.8 to 4.6 kW. The annual consumption of the active solar heating and hot-water system was lowered by close to 90 percent. The individual load factor for these building uses was significantly lower for the solar configuration—0.36 for the all-electric version and 0.07 for the solar heating and hot-water version. Substantial capital and generation problems may result for an electric utility forced to provide this backup source.

Other building investments were found to lower peak demands much more relative to total kilowatt hour consumption. The use of thermal shutters in large amounts of window expanse was found to improve load factors. For example, building 14, the masonry building, with 80 percent south glass and a winter/summer shutter, lowered peak demands to 2.6 kW and raised the load factor to 0.45. In most cases, energy-conservation investment also raised or maintained the value for the load factor.

Notes

1. The predicted cooling loads in this book are lower than sometimes experienced by actual consumers. The modeling of cooling loads is more difficult than that of heating loads because of the existence of latent heat. ASHRAE (1977) suggests that results be multiplied by 1.3 to account for latent heat. Some problems have also been discovered in TRNSYS in the use of a constant groundwater temperature and its effect on basement temperature. However, the most significant difference between cooling loads here and those sometimes experienced is the higher thermostat settings and air-conditioner COPs used here. Consumer acceptance of higher thermostat settings clearly represents the most cost-effective alternative available. For individuals unwilling to accept more extreme thermostat settings, the cost-effectiveness of energy conservation and solar energy investments will be improved.

2. It should be noted that electricity costs and rates in Colorado Springs are among the nation's lowest. Evidence of this is in Feldman and Anderson (1976a, p. 58) where only one utility, Sacramento Municipal Utility District, had lower costs. These lower rates extend many facets of the analysis to comparisons of other lower-price fuel alternatives, such as natural gas and oil.

3. See, for example, S. Feldman, "A Marginal Cost Analysis for Gas Utilities," Russell Sage Foundation Discussion Paper No. 5, New York, 1978.

5

Policy: The Impact of Utility Rates, Solar Tax Credits to Consumers, and Utility Financing of Solar Energy Systems

The impact of federal, state, or local policies on solar customers can be assessed by using the model and methodology developed in previous chapters. The methodology can be extended to incorporate other solar technologies such as photovoltaics, methane generation, and wind energy power on a decentralized basis. (A number of models describe the photovoltaics-utility interface. See Flaim et al. 1979). Energy-conservation measures and passive solar buildings are becoming rapidly commercialized, and examples of the policy impacts on these technologies provide not only timely relevance but needed evaluation for establishing precedents. (See, for an example of extending the model to account for the "sell-back" of power, Feldman and Breese 1978.) While a large number of possible policy options exist in order to accelerate the commercialization of solar technologies (for example, Bezdek et al. 1977; Miller and Thompson 1977; Feldman and Berz 1979; three major sets of economic policies stand at the vanguard: the impacts of utility rates, tax credits at the federal and state levels to solar users, and utility financing of solar applications. In order to demonstrate these impacts, the model was applied to a large number of utilities, and the results serve as illustrative applications on which the analysis is based.

Observations on Utility Rate Structures and Decentralized Solar Energy Consumers

As a result of the increasing market penetration of solar hot water and space conditioning of buildings using electric backup, a number of central issues have arisen with regard to the appropriate rate schedule for solar customers. This section reviews the state of the art of the major analytical problems associated with rate structure impacts and rate composition for solar consumers. (The legal implications of various rates for solar customers are reviewed in Feuerstein 1979. In addition, that author discusses the role of the National Energy Act in rate reform for solar customers.)

Some twenty-three state public-utility commissions have considered some form of rate structure for solar users, and the federal Public Utility Regulatory Policy Act voluntary guidelines have just been issued for public review. In most cases the rates are promotional. However, as will be shown, some of those touted as promotional are indeed inhibitory and depart from sound rate designs. The task of rate setting and rate approval is marred by the uncertainties surrounding the analysis of the impact of solar users on utility load curves and finances. Another consideration which has not been satisfactorily addressed is the definition of a solar customer according to a set of clearly identifiable technical characteristics. The subsequent review and analysis of the literature in the context of rate setting shows the major findings of a number of studies and will lend insight into some major solar-rate issues.

Rate Setting

For convenience let us define three major inputs into the rate-setting process: (1) customer class definition and load patterns; (2) costs and class revenue responsibility; and (3) rate structure design principles.

Customer Class Definition and Load Patterns. One must first attempt to show that in setting rates for a particular customer class, that customer class has a distinctly different set of load patterns from other customer classes. In the case of solar energy systems, this needs to be shown. A basic condition underlying the solar technologies concerns their dependency on climatic and weather conditions. [A number of solar technologies are not climate-dependent in energy production (see, for example, Feldman and Breese 1978), but for the moment we will not discuss these.] The evidence that solar customers in the case of space-heating and hot-water technologies have different load patterns from their conventional (heat pumps or electric furnace/baseboard) counterparts is not completely conclusive. For example, Feldman and Anderson (1976a) have shown that the load patterns of active solar heating systems and cooling customers are different from those of their conventional buildings. Kelly et al. (1977) have found similar results for solar electric and solar space conditioning. Lorsch (1977) has found that load factors of solar buildings are significantly lower than those of conventional buildings. Asbury, Maslowski, and Mueller (1979) indicate the same findings as in the Lorsch report with regard to winter space heating. At least two of the studies cited, Feldman and Anderson, and Kelly, do state, however, that solar impacts on load patterns are more a function of geography, solar system type, and sizing and that a generalization on the impact of solar on load patterns can hardly be made between utilities and even within larger utilities which service diverse geographical zones.

This brings up a variety of questions concerning solar customer class definitions. First, does the variance in load pattern demand and load factor of solar buildings change so insignificantly with solar system design and sizing? Is the change smaller than the variance between systems of different nonsolar design (buildings and their associated appliances)? There is evidence in the case of space heating and hot-water heating that this is true, particularly in comparison to a set of building envelopes that vary from insulated to well insulated. For example, tables 5-1 to 5-3 show the kilowatt demand at utility peak for three utilities located in Texas. The tables demonstrate, in large part, gross variations in building demand with modifications of: building envelopes; various heating, cooling, and hot-water heating appliances; and fenestration. The methodology used to demonstrate this effect is outlined in previous chapters. It is quite evident that variations among solar buildings at the time of utility peak are less than the variation in utility peak demands among nonsolar buildings. Similarly, tables 5-4 to 5-6 show that in examination of the kilowatt hour consumption of buildings, different HVAC systems, fenestration, and building envelopes produce enormous variations. Once again, nonsolar variations are greater than solar variations. Algebraically, it was shown in chapter 3 that if the variance of solar demands is well within the variance of the prediction of conventional peak demands, then the impact of solar has less relevance in forecasting for capacity expansion. The dynamics of the interaction of the building's fenestration and envelope with solar system's performance are complex; therefore one should not assume simple linearities between buildings from looking at the tables. Coincident load factors in some of those buildings that had higher energy-conservation measures dropped below those of less insulated buildings and were lower than those of some solar buildings. (Coincident load factors are defined as a building's average annual demand divided by the coincident peak demands with the utility.) Under these cases, at least for the Texas utilities examined, it would be difficult to define a solar customer. This is especially illustrative since some of the building mass/fenestration combinations are representative of "passive designs (albeit, suboptimal). Generally, specification of solar rates would be extremely difficult and more than likely unjustified, particularly if a series of rates were not also specified that were dependent on building envelope characteristics.

If one were to define the customer class/subclass more narrowly to encompass a particular appliance that has fewer of the properties associated with complete space heating dynamics, such as hot-water heating or thermal energy storage using off-peak electricity, the problem of customer class definition would be less aggravating. Solar hot-water systems, for example, are easier to model and their performance is simpler to simulate; thus they are easier to classify for rate-making purposes. In an analysis of two California utilities, Feldman et al. (1979) have shown that over a five-year period of examining utility peak demands, solar hot-water heater demands were 0 to 30 percent of the peak demands of conventional hot-water heaters, with collector size not being a highly sensitive variable.

Table 5–1
Impact of HVAC Systems on Peak Thermal Performance

	Kilowatt Demand at Utility Peak[a]		
Building Description	Utility 1	Utility 2	Utility 3
Base building, without collector	7.9	8.3	7.3
Base building, with 17.1-m^2 liquid collector	7.6	8.0	7.2
Base building, with 40-m^2 air collector	7.6	8.1	7.2
Energy-conservation building, without collector, without heat pump	4.6	4.8	4.1
Energy-conservation building, with 17.1-m^2 liquid collector, without heat pump	4.3	4.5	4.0
Energy-conservation building, with 3-ton air-to-air heat pump, without collector	4.0	4.2	3.5
Energy-conservation building, with 3-ton water-to-air heat pump, without collector	3.7	3.7	3.3
New-standards building, without heat pump, without collector	5.1	5.3	4.7
New-standards building, without heat pump, with 17.1-m^2 liquid collector	4.8	5.0	4.5
New-standards building, without collector with 3-ton air-to-air pump	4.2	4.6	3.8

[a] Summer-peaking utilities located in Texas.

Table 5–2
Impact of Building Envelope on Peak Thermal Performance

	Kilowatt Demand at Utility Peak[a]		
Building Description	Utility 1	Utility 2	Utility 3
Base building, R4 walls, R11 ceiling, low thermal mass	7.9	8.3	7.3
New-standards buildings, R11 walls, R19 ceiling, low thermal mass	5.1	5.3	4.7
Energy-conservation building, R19 walls, R38 ceiling, low thermal mass	4.6	4.8	4.1
Masonry building, R20 walls, R38 ceiling, high thermal mass	4.8	4.9	4.1
Masonry building, R20 walls, R38 ceiling, high thermal mass, 45% south glass, double glazing, radiation-sensitive shutters	2.8	2.7	2.3
Energy-conservation building, R19 walls, R38 ceiling, medium thermal mass, 45% south glass, double glazing, time-of-day controlled shutters	2.6	2.6	2.2

[a] Summer-peaking utilities located in Texas.

Table 5-3

Impact of Fenestration on Peak Thermal Performance for a Utility in Texas

Building Description	Kilowatt Demand at Utility Peak[a]		
	Utility 1	Utility 2	Utility 3
Base building, 22% south wall glass, single glazing, without shade	7.9	8.3	7.3
Base building, 9% south glass, single glazing, without shade	7.2	7.7	7.2
Base building, 45% south glass, single glazing, without shade	8.1	8.7	7.2
Base building, 22% south glass, double glazing, without shade	7.4	7.8	6.8
Base building, 22% south glass, single glazing, 1.22-m shade	7.5	7.9	6.7
Masonry building, 45% south glass, single glazing, without shade, shutters	5.0	5.1	4.6
Masonry building, 45% south glass, single glazing, 1.22-m² shade, without shutters	4.4	4.3	3.9
Masonry building, 45% south glass, single glazing, without shade, radiation-sensitive shutters	2.8	2.7	2.3
Masonry building, 45% south glass, single glazing, without shade, time-of-day controlled shutters	3.0	2.9	2.5
Energy-conservation building, single glazing, without shutters, 22% south glass, low thermal mass	4.6	4.8	4.1
Energy-conservation building, double glazing, time-of-day controlled shutters, 45% south glass, medium thermal mass	2.6	2.6	2.2

[a] Summer-peaking utilities located in Texas.

Table 5-4

Impact of Fenestration on Annual Thermal Performance for a Utility in Texas

Building Description	Auxiliary Demand (kWh)
Base building, 22% south glass, single glazing, without shade	42,961
Base building, 9% south glass, single glazing, without shade	41,249
Base building, 45% south glass, single glazing, without shade	42,667
Base building, 22% south glass, double glazing, without shade	39,705
Base building, 22% south glass, single glazing, 1.22-m shade	42,732

Table 5-4 continued.

Building Description	Auxiliary Demand (kWh)
Masonry building, 45% south glass, single glazing, without shade, without shutters	24,734
Masonry building, 45% south glass, single glazing, 1.22-m shade, without shutters	24,153
Masonry building, 45% south glass, single glazing, without shade, radiation-sensitive shutters	19,086
Masonry building, 45% south glass, single glazing, without shade, time-of-day controlled shutters	18,969
Energy-conservation building, single glazing, without shutters, 22% south glass, low thermal mass	22,564
Energy-conservation building, double glazing, time-of-day controlled shutters, 45% south glass, medium thermal mass	15,092

Table 5-5

Impact of HVAC Systems on Annual Thermal Performance for a Utility in Texas

Building Description	Auxiliary Demand (kWh)
Base building, without collector	42,961
Base building, with 17.1-m^2 liquid collector (19° tilt)	36,390
Base building, with 40-m^2 air collector (19° tilt)	34,794
Base building, with 17.1-m^2 liquid collector (30° tilt)	35,927
Energy-conservation building, without collector, without heat pump	22,564
Energy-conservation building, without heat pump, with 17.1-m^2 liquid collector (19° tilt)	17,058
Energy-conservation building, without collector, with 3-ton air-to-air heat pump	17,509
Energy-conservation building, with 3-ton water-to-air heat pump, without collector	19,199
New-standards building, without heat pump, without collector	23,823
New-standards building, without heat pump, with 17.1-m^2 liquid collector	20,013
New-standards building, with 3-ton air-to-air heat pump, without collector	19,978

If, indeed, decentralized solar *consumers* (we do not refer to cogenerators, who are clearly more identifiable) are not a distinct class of consumers by load-pattern criteria, then one should approach the pricing problem in a similar manner to the foundations of the federal Public Utility Regulatory Policy Act voluntary guidelines for solar customer: *In general, price backup*

Table 5-6

Impact of Building Envelope on Annual Thermal Performance for a Utility in Texas

Building Description	Auxiliary Demand (kWh)
Base building, R4 walls, R11 ceilings, low thermal mass	42,961
New-standards building, R11 walls, R19 ceiling, low thermal mass	25,823
Energy-conservation building, R19 walls, R38 ceiling, low thermal mass	22,564
Masonry building, R20 walls, R38 ceiling, high mass	24,860
Masonry building, R20 walls, R38 ceiling, high thermal mass, 45% south glass, double glazing, radiation-sensitive shutters	18,819
Energy-conservation building, R19 walls, R38 ceiling, medium thermal mass, 45% south glass, double glazing, time-of-day controlled shutters	15,092

energy at marginal cost without worry as to their particular impact on technologies (customer classes in particular circumstances).[1] In the long run this is more desirable. Departures from this condition should be the exception rather than the rule.

Costs and Class Revenue Responsibility. If major difficulties are evident in class identification, cost allocation becomes fraught with the same drawbacks. However, a number of studies have attempted analyses which yield some inference on the costs of providing backup power for solar technologies. It must be kept in mind that the impact of an individual's load must be evaluated with reference to the cost to serve that individual. Yet, rate schedules are created not to produce equivalent profits from each kilowatt hour sold, but to combine broad ranges of users into classes. In an initial study by Bright and Davitian (1978), it was found in the case of Long Island Lighting Company (LILCO) that the estimated marginal cost of electricity for backup to solar hot-water systems was less than the average cost for the period of the mid- to late-1990s. Bright and Davitian have also found this to be the case for space heating for LILCO and Public Service of New Mexico.[2] This simulation used the assumption that long-run marginal costs could be calculated through the use of a production cost model, WASP II. Lorsch (1977), using two utility accounting schemes for two respective utilities in Pennsylvania, concluded that solar space-conditioning systems would have higher backup costs than their conventional counterparts. Kelly et al. (1977), using a variant marginal-cost scheme, have shown that the cost of providing backup power for solar technologies generally varies from case to case.[3] Their analysis shows that backup energy for solar devices need not cost utilities more than energy to conventional homes and, in fact, may cost less. These findings are similar to but not entirely consistent with those found in Feldman and Anderson (1976a). In that study utility costs are highly variant for different solar customers, depending on system design and sizing. Using the Cicchetti, Gillen, and Smolensky (1976) model for estimating long-run marginal costs, they found that in several utilities solar customers' backup costs would be less than the cost to serve conventional buildings per kilowatt hour consumed. This did not hold true in other utilities that were examined.

Obviously, the cost to provide service to solar cogenerators is going to be sufficiently different depending on the technologies employed and their sizing as well as the sell-back load schedule. Kelly et al. (1977) have shown that the cost of providing backup power to solar electric houses which generate electricity is reduced if the homes are not permitted to sell electricity to the utility. Feldman and Breese (1978) show the same situation to prevail in the case of waste-produced methane gas. By using the Cicchetti, Gillen, and Smolensky marginal-cost methodology to calculate the sell-back rate, utility savings can be seen to be quite considerable when the sell-back of electricity occurs.

Revenue responsibility from solar customers under various rate schedules is a highly sensitive issue, especially in light of the findings expressed thus far. Lorsch (1977) found that under present rate schedules the utility could not recover costs incurred by solar customers. This impact of solar buildings on utilities and the subsequent impact of the rates on solar cost-effectiveness have led to discussion of the "correct" solar tariff. The following discussion focuses on these issues.

Rate Design Principles and Decentralized Solar Technologies. There has been a paucity of work in the area of examining the impact of solar rate schedules, or conventional rate schedules, on solar users and their respective utility. This is a short review of the state of the art of particular rate structures with their solar user effects.

Declining Block-Rate Schedules. Lower rates for each incremental block of kilowatt hour consumption result when utilities attempt to recover customer costs by having demand (capacity) costs in the initial consumption blocks. A solar user would pay considerably more per kilowatt hour of consumption than his conventional counterpart. This may or may not be justified, but this condition provides a disincentive to solar customers. Chapter 4 showed that the subsidization of solar customers by nonsolar customers can occur; but, depending on load patterns produced by solar customers, the opposite effect may also appear. As long as one maintains the position that the solar impact is highly variant depending on conditions mentioned previously, generalizations about rate structure impacts cannot be valid. In an example of a solar space-conditioned building in Albuquerque, New Mexico, we have shown that the solar building is subsidized by nonsolar customers. In the case examined, the solar building had 39 percent of the kilowatt hour consumption of the conventional building, but twice the kilowatt demand during the utility's peak period. Feldman and Anderson (1976a) have shown that under this rate schedule there is little incentive to optimize the sizing and design of solar space-conditioning systems.

Inverted Rates. The inverse of the declining block-rate schedule—inverted rates—would obviously lower the average kilowatt hour cost to solar consumers. There has been no systematic inquiry as to the impact of such rates on solar customers. A form of an inverted rate schedule has been proposed by Pacific Gas and Electric (PG&E), combined with a time-of-use tariff.[4] The off-peak period is coupled with a lifeline rate and is, in essence, an inverted schedule. This rate was in part a response to criticism by Feldman et al. (1979) who claimed that the existing time-of-use rate schedule for solar customers encouraged thermal energy storage, which for PG&E, an "energy-constrained" utility, is a cost-ineffective technology.

Demand or Demand/Energy Rates (Hopkinson Rates). This rate schedule is becoming more popular in the residential sector, whereas before it was primarily used in the commercial and industrial sectors. This rate schedule was designed to increase customers' load factors. The rate schedule makes no provision for the assessment of individual demand at the utility's peak demand. The tariff is linked to the individual's peak, regardless of whether it coincides with system peak, so it is not relevant to the question of whether the solar energy consumer will require supplemental power during the system peak. This is a basic shortcoming of this tariff. It may even have the effect of shifting weather-sensitive loads to system peak. Dickson, Eichen, and Feldman (1977) discuss this tariff and its impact on solar customers at greater lengths. They report that in a study of the Colorado Springs municipal utility, a Hopkinson rate that was proposed discriminates against the solar energy user. For the solar building the maximum monthly demand occurred in off-peak hours with the exception of one month during the year. It was suggested that the utility or consumer exercise some sort of load management control during this one month.

Time-of-Use or Time-of-Day Pricing Based on Marginal Cost. Almost no empirical or simulated work exists or the impact of time-of-use pricing on solar customers. If one is to obey standard economic postulates by equating the long-run marginal costs of solar and energy conservation with the long-run marginal costs of backup energy, then some benchmarks need to be set.

For many years, economic opinion held that the appropriate theoretical benchmark for utility pricing was long-run incremental cost (LRIC). The transition from theory to practice, however, resulted in a great deal of regrettable confusion in regulatory circles.[5] Rather than chronicle the unfortunate turns that discussion has taken, in the following pages a variety of approaches to marginal-cost analysis are described which are consistent with the rather general notions one draws from theory, but which also are quite pragmatic in attempting to meet the requirements of those involved in the regulatory process.

There is an important distinction between the precise economic terms *short-run marginal cost* and *long-run marginal cost*, which should be understood by regulators and other participants in regulatory proceedings. In the purest, most theoretically sound sense, the economically efficient base for pricing is short-run marginal cost. There is substantial variation for most utilities in the operating cost of providing electricity in different periods. These operating costs are a very significant component of short-run marginal cost.

There is, however, a second component of short-run marginal cost. It is the incremental cost created by having demand at a particular time exceed the utility's available capacity. In a purely theoretical sense this can be measured by determining the cost consequences related to certain users of electricity having their service interrupted for various lengths of time. Such costs will depend on the duration and distribution of power outages.

In practice, it is very difficult to think how one might go about measuring the cost consequences of demand exceeding supply. Determining a "proxy" for this cost is a useful way of approaching the problem. One proxy which is very much available and understood by utility commissions and utility firms is how much a utility must spend in order to provide additional generation, transmission, and distribution capacity sufficient to avoid (or at least reduce) the probability of demand exceeding available supply.

This additional cost consideration can be characterized as the marginal generating and transmission capacity cost. But these costs will not be incurred in the short run. In the case of generation capacity, they will be incurred over several years. Transmission and distribution capacity cost will be incurred over a period from several months to several years. These marginal or incremental capacity costs can be thought of as long-run marginal costs. To compare both the short-run, or marginal, operating cost and the long-run marginal generation, transmission, and distribution capacity cost, it is necessary to recognize that the latter is serving as a proxy for the short-run cost of an outage. This is a pragmatic way of bringing together the theoretically pure and practically sound bases for pricing electricity by using the concept of marginal cost. It uses information that is maintained and used by utility dispatchers and long-range system planners and translates these calculations into a common basis for establishing price signals.

This discussion is a heuristic treatment of this subject for nontheorists. However, independent theoretical work by both Boiteux (1956) and Turvey (1968) has demonstrated the equivalence of these short-run pricing rules. Specifically, Boiteux has shown how depreciation, in economic terms rather than accounting terms, is the link between the long-run and short-run cost bases for pricing. Turvey utilizes a similar method, but he bases his rules of pricing on both a short-run rule, which uses the marginal energy cost and the cost involved in restricting demand to whatever level of supply is available, and a long-run investment rule, which utilizes the cost-minimizing approaches of the system engineer as the formal basis for determining pricing and tariff structures.

Needless debate on short-run versus long-run marginal-cost pricing should be avoided by all regulatory commissions. Both methods can, after all, be reduced to a set of equivalent tariff recommendations. It is even more important to avoid the misconception that LRIC is some magic method, justified by economics for increasing price levels or revenue requirements. Economics does not require such a result, even if the term *LRIC* is invoked in the explanation. If a utility commission believes earnings should be increased because of changed conditions, rising costs, or any other matter, it should not base its decision on the form of pricing reform that goes under the heading of marginal-cost pricing. Such revisions in revenue requirements or earnings must stand on their own merits and should not be included as part of a package for reforming or restructuring electricity tariffs.

An Econometric Approach. There are two general approaches to marginal-cost estimates. One is based on an econometric approach for determining marginal cost. To understand this approach, it is useful to review the chain of economic theory that results in a marginal-cost pricing schedule. Economists start out with a conviction that a firm arranges its inputs and outputs in some technological fashion which is called a production function. The production function and the prices or unit cost of the various inputs are selected in order to minimize the cost of various levels of output. The schedule that describes the cost-minimizing conditions for each output level is called a marginal-cost schedule. It is also called a supply schedule.

Some analysts believe that since economic theory is based on the production-function approach to describe the manner in which a theoretical marginal-cost schedule will be determined, this approach should be followed by those who seek to measure marginal cost for various electric utilities. The econometric approach gathers information concerning the various types of output and the various forms of input that are used to produce electricity in different years for the same utility as well as by different utilities in the same year. This information can be statistically analyzed to estimate a production function for a typical electric utility by using various econometric estimation techniques.

By taking the characteristics of a particular utility with respect to the unit cost of the various inputs as well as the output characteristics of that particular utility in terms of the services that must be supplied, the econometric approach may be used to estimate marginal cost. In other words, the statistically estimated general production function is used to estimate the marginal cost of providing electricity for a particular utility and its various service categories.

The production-function econometric approach for evaluating marginal cost is only as good as the correspondence of the general production function to the actual decision-making function of the specific utility being studied. It is the traditional approach used by econometricians to quantify economic functions, but it may not be the best way of estimating the actual decision framework for a particular utility. Its greatest strength is that the approach is almost entirely objective because different analysts considering the same set of data are very likely to produce a similar production function and, therefore, the same marginal costs. Its greatest weakness is that engineering data on the costs of producing electricity are generally better for the time pattern of outputs of a specific utility than for the industry as a whole.

The production-function approach either must make assumptions about the time pattern of cost and output, in order to determine marginal cost, or must be restricted to serving as a device for allocating the level of marginal cost to the various customer categories of a utility. This is necessary because information on sales or outputs at the various voltage levels is generally available, but variations by time of use are not. As a theoretical as well as practical variant to average cost allocation by customer categories, the production-function approach has great potential. As a device for calculating marginal cost at different periods, however, either the approach is dependent on assumption, rather than observation, or it requires access to data which are not generally available.

An Engineering Approach. The engineering approach for marginal-cost evaluation is based on a detailed analysis of the actual cost and decision-making framework of a specific utility. Electric utilities practice cost minimization in two important ways. The system dispatcher, using sophisticated, computerized analysis, determines the least-cost way of supplying any given load at any time for the various locations throughout the service area. This requires a detailed knowledge of the marginal operating cost of each plant in the system, but all utilities practice this procedure to a large extent. The engineering approach is based on the exact information used to minimize operating cost, instead of on production functions gathered from utilities which might have quite different characteristics. This approach is not entirely devoid of statistical analyses, but the data come from the actual experience of a single utility or single generating pool. It is a description of what engineers expect to happen.

The second type of cost minimization that takes place within a specific utility is related to the ways in which a utility minimizes the cost of expanding its generation, transmission, and distribution capacity. Once again, utilities generally use very sophisticated methods of analysis to determine the various costs of meeting future capacity requirements. Cost minimization is the objective that determines which of the various plans will be selected. The cost-minimization information inherent in a specific expansion plan provides a direct estimate of the marginal-capacity cost of generation, transmission, and distribution. The engineering approach uses the actual cost-minimization calculations determined by the system dispatchers and system engineers of a particular utility system.

The main criticism of the engineering approach is that the analysts must have confidence in both the data provided and the engineers who operate and plan the specific utility system. In the current climate of debate before regulatory commissions, using a system which is so highly dependent on what many believe to be the subjective judgment of employees of the utility being regulated poses an obvious problem.

This problem is not particularly new to regulation, because all information for determining the total cost and rate base of a utility is subject to the same sort of potential bias. It is new, however, to place this type of dependence on the utility planning and operating engineers. However, having the system planners and operating engineers become more intimately involved in tariff and marketing questions may produce a greater acceptance by the utility of tariff design and load management, rather than reliance on new capital expansion, as the appropriate response to continued growth in electricity consumption during peak periods.

The engineering approach is the approach most directly available for calculating marginal cost on a time-of-use and voltage-varying basis. The econometric production-function approach may provide useful complementary information, but it has yet demonstrated that it is also capable of providing a time-of-use and voltage-varying basis for establishing tariffs. Comparison of these two approaches should continue, but the time of regulators and the records of regulatory proceedings should not be dominated by such a debate.

Regarding the engineering approach, several particular methods have been suggested by various consultants. First, some assume that a utility would always meet an increase in demand by building a low-capital-cost, high-operating-cost peaking plant. This approach would always overestimate the marginal cost of generation if a utility could supply an increase in demand in a less costly manner. If acquiring such a facility were the least-cost strategy, this approach would indicate the correct answer; if not, it would yield an overstated approximation.

A second approach sometimes recommended for calculating marginal generating cost assumes that if there were an unexpected increase in demand, the utility would completely and totally reoptimize its entire system to take into account this uncertain and unplanned demand. It quite unrealistically assumes that the entire utility plant would be made to respond to this new circumstance. It assumes that the utility can dismantle its old and already acquired plant and equipment and purchase at current cost a completely new system that would provide sufficient electricity to meet this unexpected increase in demand in the least-cost or optimal fashion. Since the utility would certainly prefer to be free of the constraints mandated by past decisions based on different projections of future demand, the reoptimization approach for evaluating marginal cost will always produce an estimate of the marginal generating cost less than or equal to the actual marginal generating cost that would be experienced in reality. Because the analyst who uses the reoptimization approach ignores real-world constraints, there will always be a tendency to underestimate the actual cost a utility would incur.

There is a third approach for calculating marginal generating cost, which can be characterized, for lack of a better term, as the "what if" method. Under this approach, a utility system engineer is asked how the utility would actually respond to an unexpected increase or decrease in demand during various periods. In other words, by asking the "what if" question, one can determine how the utility would respond to the change in demand and still minimize cost. The answer will indicate the actual marginal generating cost for the utility. This approach is unquestionably the most subjective of all. But the utility system engineer's response is the most accurate point estimate of marginal generating cost.

There are, of course, several ways in which a utility may respond to the circumstances of changing demands, either increasing or decreasing, during peak periods. By understanding these various methods, one can get a better understanding of what this "what if" approach is all about and, further, what the engineering estimation approach to marginal cost is likely to mean in practice.

A frequently discussed form of the "what if" engineering approach for estimating marginal generating cost assumes that the utility will meet an increase in demand during peak periods by moving ahead of schedule (bringing forward) the next generating plant it had planned to bring on-line, presuming that there is still some discretion concerning the precise timing of that plant. The opposite is presumed for a decrease in demand. The marginal generating-capacity cost would include the extra carrying changes associated with having that plant built and operating ahead of schedule. It would also include any reduction in annual system operating costs that would result from bringing such a plant on-line ahead of schedule.

If, on the other hand, demand were expected to decrease, then this method of changing the time pattern of future expansion plans would examine the reduced carrying cost of postponing the acquisition of a future plant (adjusted for any higher fuel costs associated with such a postponement).

If moving plans forward or backward in the planned construction schedule is the way in which a utility would actually respond to a changed demand during peak periods, then this method is the appropriate approach to use for calculating marginal generating cost. However, if a utility would not respond in this manner or is incapable of responding by changing the time of its future plant acquisitions, then an alternative approach is in order.

A utility might also respond to increases in demand by increasing its purchase of power and/or energy from neighboring utility systems to meet its requirements. If this is the least-cost way to respond to the increase in demand and, indeed, if it is possible to purchase increased quantities to meet demand, then the price of the purchase power is the appropriate measure of marginal cost.

In distinguishing peak and off-peak periods for tariff purposes, it is first necessary to determine whether the variation in costs among time periods is sufficiently different that it makes economic sense to establish different prices for the electricity supplied in each period.

The establishment of peak periods is based on two criteria: (1) the load or demand expected in a particualr period and (2) the amount of expected generating capacity. There are several methods for analyzing the relationship between these two criteria. There are differences in both the information that may be available to a particular utility and the level of sophistication of the analysis.

One method of analysis is called the *loss-of-load probability method.* The relationship of expected demand to supply will result in varying probabilities of a loss-of-load (outage) in different periods. By establishing an upper limit on the acceptable probability of a loss-of-load, it is possible to determine in which period the loss-of-load probability is expected to be greater than that limit. Such periods will be peak periods.

A second method for establishing peak periods is based on a much handier rule-of-thumb and, for that reason, is more likely to be available for all utilities. Under this method, some arbitrary percentage of reserves to available capacity is established. As a policy rule it becomes necessary to maintain this particular percentage of reserves. If there is any likelihood that demand may exceed supply and require the use of reserves, the periods in which this might happen are classified as peak periods.

The imprecision of all these approaches is apparent. However, the number of hours of peak and the related number of kilowatt hours of electricity sold during peak will have a very significant quantitative effect on actual tariffs derived from marginal-cost analysis, because a large share of the capacity cost of the utility will be recovered from the sale of electricity during peak periods. The smaller the number of peak sales, the higher the unit price of peak electricity must be. A broader definition of peak, which characterizes more hours and thus more sales as peak, would lower the estimated marginal cost of peak-period electricity and therefore also the peak-period price. (Further research into the various methods of determining the peak periods in a more objective manner will undoubtedly produce benefits greater than the cost involved. To the extent that costs significantly change with peak-period definition, this issue remains salient.)

Several other factors will affect the setting of peak periods far beyond any more precise mathematical definition of those periods. One of these is related to the economics of metering electricity. It is necessary in establishing tariffs to have a metering system for which the cost savings to the utility and/or society exceed the extra cost of metering. The greater the number of peak periods, the more complicated the necessary metering system and thus the greater likelyhood that more costly meters would be required.

It must also be remembered that the definition of peak periods for generating may not precisely coincide for specific utilities with the definition of the peak periods for the transmission and distribution systems. If the cost of transmission and distribution is sufficiently large relative to the marginal generating-capacity cost, this distinction may prove to be very significant in the establishment of prices based on marginal cost. In general, a widely dispersed utility system may find that serving demand in a particular area is constrained not by the overall ability of the system to generate sufficient power at a given time, but by the ability of the utility to deliver it to a particular point. This fact, if sufficiently important, may be reflected in higher distribution charges or in a pattern of prices somewhat different from those which exist throughout the remainder of the system.

The Problem of Second-Best. A basic tenet of welfare economics is that if all the goods and services provided by a society are produced and supplied according to marginal cost, if there is perfect information among producers and consumers, and if all externalities have been internalized, then social welfare and economic efficiency will be maximized. The theory of second-best is that if these conditions are not met, then there is no general and unambiguous second-best solution whereby some commodities are priced as marginal cost and others are not. As a mathematical proposition, the theory is not disputable. It is sometimes argued that because not all externalities have been internalized, because not all producers and consumers have perfect information, and/or because not all goods in society are supplied on the basis of marginal cost, it follows that electricity should not be supplied on the basis of marginal cost.

There are two responses to this line of reasoning. The first is that the conclusion "should not" is too strong. If marginal cost is not the basis of pricing, something else must take its place. We do not know of any obvious approach. Furthermore, cost minimization suggests marginal-cost pricing. It is not necessary for a utility or a regulatory commission to establish social welfare maximization or economic efficiency as its guiding criterion in order to design tariffs based on marginal cost. Instead, a commission may want to maximize the sales of electricity in an area for a given investment. Or it may want to minimize the cost of supplying the various loads required by the customers of a utility. Each of these two objectives suggests the utility should price electricity on the basis of marginal cost. Since both objectives are consistent with various economic, environmental, and energy goals, marginal-cost pricing should stand on its own merits without absolute reliance on the criterion of social-welfare maximization.

Second, when it comes to the matter of the possible effect of the pricing of electricity on goods and services, such as insulation or the utilization of alternative fuels, these externalities deserve to be factored into the tariffs that are approved by a regulatory commission. Such considerations should take place after the marginal-cost structure has been determined. If marginal cost should be the basis for determining the price structure, then there should be an adjustment for the revenue constraint required by regulation; and, finally, on an individual issue-by-issue basis, externalities or interdependent effects on other markets should be factored into consideration at the tariff design stage.

In other words, tariff design is a three-step process which includes (1) an analysis of the marginal-cost structure, (2) a determination of earnings and revenue requirement by using traditional techniques, and (3) an assessment of other factors which public policy deems appropriate. The actual tariff designs adopted should reflect a commonsense marriage of these separate considerations. The third comes under the heading of "second-best." Such factors as the following might be considered:

The nonrenewability of fossil fuels

The balance-of-payments impact of fossil-fuel use and the Btu equivalent (or opportunity cost) suggested by world energy prices

Pollution externalities

Site- and transmission-related externalities

Fossil-fuel production and transportation externalities

The use of an interest rate to represent the opportunity cost of displaced private capital

Income distribution

This list is not exhaustive. Some of these items may be better left to other policy institutions than the electric tariff-setting process.

Interruptible Rates. Interruptible rate structures for solar customers seldom have been considered. In some utility service areas, for some solar technologies, demand for peak electricity may be so infrequent that it may be feasible and efficient to offer a reduced rate to those users who, according to prearranged contract, would permit service interruption during designated or unexpected peak periods. An analysis of one utility found that savings from infrequent discomfort (several days per year) of having peak power interrupted may approach $600 per year (Feldman et al 1976a). Qualitatively, interruptible rates should not be very different in construction from time-of-use rates.

Thermal Energy Storage and Interruptible Service Contracts. Time-of-use (TOU) rates provide incentives to customers to invest in hardware that limits on-peak electricity consumption. The storage of off-peak thermal energy may provide greater cost savings to the individual and benefits to the utility than solar energy charged storage. An exploration of this, which compared a thermal energy storage (TES) system to the previously modeled systems, was performed in Colorado Springs. The control strategy for the TES system was simply to provide energy to storage via off-peak electricity until a critical tank temperature, sufficient to carry the building through the peak period, was surpassed.

The results in table 5–7 indicate that the TES system utilized 2,212 kWh more of electricity for heating (14,956 to 12,744) than the conventional resistance heating unit for the same building (base building 1). The benefits of TES, however, are that the entire electricity consumption occurred off-peak. The operation of the TES system was sized to ensure that no on-peak electricity was required. A smaller tank or lower tank temperature would lower tank losses and improve TES cost-effectiveness. Optimization of the TES system is beyond the scope of this project. The benefits to the utility in fuel savings, improved load factor, and reduced capacity requirement are intuitive. Benefits to the consumer can be realized only if some rate incentive is provided.

Table 5-7

A Comparison of Thermal Energy Storage versus Electric Resistance Heating for a Base Building, No Shade (1C), in Colorado Springs

| | Thermal Analysis | | | |
| | Kilowatt hours | | | Demand at |
	Peak	Off-Peak	Total	Utility Peak (kW)
Thermal energy storage	0	14,956	14,956	0.7
Electric resistance heating	942	11,802	12,744	5.8

| | Financial Analysis | | |
| | First-Year Analysis | | |
	Incremental Capital Cost ($)	Electric Bill— TOU Rate ($)	Total Cost to Consumer ($)
Thermal energy storage	264	379	645
Electric resistance heating	0	420	420

	Medium Life-Cycle Cost		
Thermal energy storage	3,070	11,013	14,083
Electric resistance heating	0	12,227	12,227

The results of the first-year cost analysis reveal that this particular TES system is not cost-effective to the consumer under the hypothetical TOU tariff. Although the TES system does lower costs of fuel, the reduction is insufficient to compensate for the added capital expense of thermal storage.

Despite complete avoidance of peak-period electricity charges at $0.0996/kWh (off-peak electricity was $0.0167/kWh), the TES system does not save money. The conventional electric resistance heating system in Colorado Springs utilizes a majority of its energy in the off-peak hours so that potential savings by TES is limited. Another important note is that unlike other investments explored here, the economics of the TES system becomes worse in the life-cycle analysis. Because TES systems consume more electricity, future rises in price may make them less economic with time.

The real benefits of the TES system are in the reduction in the kilowatt capacity requirement of the utility. Other strategies might also enable the utility to reduce kilowatt requirements without a commensurate increase in kilowatt hour consumption. Peaking problems can be alleviated by utilizing interruptible service contracts with electric heating customers. The interruptible service lowers the kilowatt capacity requirement of the utility. The infrequency of peak electricity demand by solar buildings may make interruptible service contracts ideal. Examination of building room temperature has revealed that large amounts of building thermal mass damp heat flows to such a degree that the exclusion of on-peak electrical heating does not appreciably alter internal room temperature or comfort.

The variation in utility rates is clearly most responsible for variations in cost-effectiveness results among utility locations. The low cost of fuel and capacity for Colorado Springs clearly favors the consumption of electricity as compared to other locations such as Worcester or Albuquerque. This phenomenon is best illustrated by examining the impact of higher electric rates on building cost-effectiveness in Colorado Springs (see table 5–8). The "appropriate block" electric rate for Colorado Springs was changed from the original $0.025 kWh to $0.035 kWh and $0.045 kWh, with the latter closely approximating the value for Worcester and Albuquerque. There is much debate as to which electric rate best represents the block of incremental heating loads to solar. We have used the lowest block rate because it is this electricity that solar investment clearly replaces.

Utilizing the original electric rates ($0.025/kWh), the all-electric base building 1C is the lowest cost alternative in both the first-year and medium life-cycle scenarios. When costs are raised by one cent to $0.035/kWh, the least-cost alternative shifts. In the first-year analysis, the all-electric energy-conservation building 2 with 80 percent south glass, thermal shutter, and shade has a lower cost than base building 1C. For the medium life-cycle analysis and a price of $0.035/kWh, the all-electric masonry building with 80 percent glass and a winter/summer differentiating shutter, was the least-cost alternative to the customer.

Utilizing the rate of $0.045/kWh, which was similar to the rate for Albuquerque and Worcester, produces significant changes. In the first-year analysis, the same energy-conservation building (6C) was still the least-cost alternative. In the medium future life-cycle scenario sixteen buildings had lower costs than the all-electric base building. The least-cost alternative was 14B, the masonry building with 80 percent glass, thermal shutters, and a solar hot-water system. If costs in Colorado Springs were as high as they are in Worcester, it is not surprising to find that the cost-effectiveness of solar heating systems would be more favorable in Colorado Springs, where weather is more favorable. That is illustrated by the large number of A-type buildings that have costs lower than base building 1C in the life-cycle analysis with $0.045/kWh as the electric rate. A further comparison with other utilities is provided in a later section of this chapter.

The Impact of Tax Credits on Solar Consumers

Conflict often exists between the immediate needs of individuals and the long-run optimum for society. The individual consumer is constrained in

Table 5–8
Sensitivity of Consumer Cost-Effectiveness to Electric Rates; Colorado Springs[a]

Rate ($/kWh)	Least-Cost Alternative	Other Buildings with Costs Lower than Base Building 1C
	First-Year Cost Analysis[b]	
0.025	All-electric base building, no shade (1C)	—
0.035	All-electric energy-conservation building 2, with shade, 80 percent glass, thermal shutters (6C)	None
0.045	All-electric energy-conservation building 2, with shade, 80 percent glass, thermal shutters (6C)	None
	Medium Future Life-Cycle Analysis[c]	
0.025	All-electric base building, no shade (1C)	—
0.035	All-electric masonry building, 80 percent glass without shutters (14C)	6C, 14B
0.045	Solar hot-water masonry building, 80 percent glass, without shutters (14B)	1A, 2A, 3AC, 4AC, 6BC 9A, 13BC, 14C, 15BC

[a] No tax credits applicable.

[b] For a description of building numbers see table 2-3.

[c] Discount rate = 0.08, interest rate = 0.085, electric rate increase = .07 per year, fuel cost increase = 0.10 per year.

energy investment by the need to invest money "up front" for potential future savings. The government has found this first cost constraint to be so significant that tax policies were developed which lessen the initial cost outlay (Opinion Research Corp. 1976). Perhaps the most significant policies to date are the tax credits which allow the investor to recoup a percentage of the initial investment in the form of a reduction in income taxes.

Four different tax policies have been examined here to assess their impact on cost-effectiveness to the consumer. In addition, the economic efficiency of these tax-credit policies will be addressed with regard to maximizing societal welfare. The policies are:

1. The proposed federal tax policy as of April 1978 for solar energy and energy conservation: for active solar, 30 percent of the first $2,000 and

20 percent of the next $8,000; for energy conservation, 20 percent of the first $2,000.[6]

2. An alternative federal tax policy where solar energy and energy-conservation investment are treated equivalently: 30 percent of the first $2,000 and 20 percent of the next $10,000.

3. The California tax-credit policy for solar energy: credits of 55 percent (up to $3,000 in credit) for active solar systems, passive systems, and energy conservation in conjunction with solar.

4. An alternative California tax policy where energy conservation not in conjunction with solar also is eligible for tax credit.

The results of the different tax policies on consumer first-year cost are shown in table 5–9. The federal tax credit for solar has very little effect on the first-year cost analysis. The all-electric base building with shade 2C is still the least-cost alternative, although the all-electric energy-conservation building with shade 3C is now less expensive than the all-electric base building with shade 2C. The reduction is enough to make the inclusion of additional insulation cost-effective between both shaded buildings even in the first year. Otherwise, the desired impact of lowering the capital costs of larger active systems still leaves the costs of buildings with active systems higher than those of energy-conservation and passive buildings. This occurs in spite of the fact that the latter two investments do not receive nearly the same benefits from the tax-credit policy.

A federal policy which treats each of these investments equally gives a definite advantage to additional insulation over a solar energy investment since building 3C has a lower cost than building 2C.

The California tax-credit policy represents broader commitment to the development of alternative investments to electricity. Tax credits are extended in larger quantities to passive designs, including overhanging shades, thermal shutters, window size, and glazing, and thermal mass. Energy conservation in conjunction with solar is eligible. Only the investment in energy conservation not in conjunction with solar is not eligible. This excludes from credit any investment in the all-electric buildings that just improve insulation levels and retrofitting the minimally insulated building with insulation.

The significant finding is that the all-electric masonry building with thermal shutters, 80 percent south glass, and no shade was the lowest cost alternative. The investment in energy conservation fell in rank, being supplanted by several passive buildings. In the California solar tax credit, where energy conservation not in conjunction with solar does not qualify, the cost-effectiveness of energy-conservation alternatives was significantly altered. Tax savings accruing to the passive and active solar investments

Table 5–9
Impact of Tax Credits on Consumer Cost-Effectiveness; Colorado Springs,
First-Year Cost Analysis

Tax Policy	Lowest Cost Alternative	Other Buildings Lower than Base Building with Shade (2C)
1. No tax policy	All-electric building with over-hanging shade (2C)	None
2. Federal solar tax policy (proposed 4/78) (Solar: 30% of first $2,000, 20% of next $8,000; conservation: 20% of first $2,000)	All-electric base building (2C) and all-electric energy-conservation building without total wall (3C)	No others
3. Federal solar = energy-conservation tax policy (*hypothetical*) (Solar and energy conservation, 30% of first $2,000 and 20% of next $10,000)	All-electric energy conservation building without total wall (3C)	No others
4. California solar tax policy (*existing*) (55% of active and passive solar and energy conservation in conjuntion with solar up to $3,000)	All-electric masonry building (14C)	13C, 15C
5. California energy conservation = solar tax policy (*hypothetical*) (same as 4 but includes all energy conservation)	All-electric masonry building (14C)	3C, 4C, 6C, 13C, 15C

All buildings are with overhanging shade and/or thermal shutters.

were not received solely by energy-conservation alternatives; however, energy-conservation investment still remained favorable in comparison to active solar heating investment.

When benefits of the tax incentives were included for energy conservation not in conjunction with solar applications, the position of these investments improved. Investments in masonry buildings 14C and 15C and energy-conservation buildings 3C, 4C, and 6C were lower than in the all-electric base building with shade 2C. The best active system was the 17.5-m² collector system. It should be pointed out again that the inclusion of both a collector and extensive insulation, as is done here, is in part a duplication of effort. It is unclear whether separate investment in active systems or insulation would make the cost-effectiveness of the 17.5-m² collection system on a base building greater than that of the all-electric energy-conservation building with total wall.

The impact of tax credits on the life-cycle analysis reveals a clearer case of cost-effectiveness of the all-electric version of the masonry building with winter/summer shutter (14C). Under the medium life-cycle scenario (see table 5–10), it is illustrated that building 14C is the least-cost alternative for three of the four tax-credit alternatives. The existing federal tax policy for solar does not provide enough benefits from energy conservation and passive solar to change the cost advantage of the base building without shade. With a tax policy that includes benefits of passive investment (cases 3, 4, and 5) building 14C becomes the least-cost alternative.

An interesting change occurs when the high future-oriented life-cycle scenario becomes great enough to justify the inclusion of a solar hot-water system in the masonry building with winter/summer shutters. In addition, nearly all building with additional investment in either energy conservation or active or passive solar energy was less expensive than the base building 1C.

Table 5–10
Impact of Tax Credits on Consumer Cost-Effectiveness in Colorado Springs; Life-Cycle Cost Analysis, Medium Future Scenario[a]

Tax Policy	Lowest Cost Alternative	Other Buildings Lower than Base Building 1C
1. No tax policy	All-electric base building without shade (1C)	None
2. Federal solar tax policy (proposed 4/78) Solar, 30% of first $2,000 and 20% of next $8,000; energy conservation 20% of first $2,000)	All-electric base building without shade (1C)	None
3. U.S. energy conservation-solar tax policy (*hypothetical*) (Solar and energy conservation, 30% of first $2,000 and 20% of next $10,000)	All-electric masonry building, 80% glass without shutters (14C)	8C, 15C
4. California solar tax policy (*existing*) (55% of active and passive solar and energy conservation in conjunction with solar up to $3,000)	14C	6BC, 9A, 12C, 13BC, 14B, 15BC
5. California energy conservation = solar tax policy (*hypothetical*) (same as 4 but includes all energy conservation)	14C	3BC, 4BC, 6BC, 7BC, 8BC, 9A, 9B,C, 12C, 13BC, 14B, 15BC.

[a] Life cycle is 20 yr, discount rate = 0.08, interest rate = 0.085, downpayment = 20%, rate of fuel cost increase = 0.1, rate of electricity price increase = 0.07.

Table 5-11
Impact of Tax Credits on Consumer Cost-Effectiveness in Colorado
Springs; Life-Cycle Cost Analysis, High Future Scenario[a]

Tax Policy	Lowest Cost Alternative	Other Buildings Lower than 1C
1. No tax credit	All-electric masonry building 14C	6C, 13C, 14B, 15BC
2. Federal solar tax policy (proposed 4/78) (Solar, 30% of first $2,000 and 20% of next $8,000; energy conservation, 20% of first $2,000)	Solar hot-water system on 80% glass, without shutters (14B)	3AC, 4AC, 6BC, 7ABC, 8ABC, 9ABC, 13BC, 14C, 15BC
3. U.S. energy conservation = solar tax policy (*hypothetical*) (Solar and energy conservation, 30% of first $2,000 and 20% of next $10,000)	14B	3AC, 4AC, 6BC, 7ABC, 8ABC, 9ABC, 12BC, 13BC, 14C, 15BC
4. California solar tax policy (*existing*) (55% of active and passive solar and energy conservation in conjunction with solar up to $3,000)	14B	3ABC, 4ABC, 6BC, 7ABC, 8ABC 9ABC, 12BC, 13BC, 14C, 15BC
5. California energy conservation = solar tax policy (*hypothetical*) (same as 4 but includes all energy conservation)	14B	3ABC, 4ABC, 6BC, 7ABC, 8ABC, 9ABC, 12BC, 13BC, 14C, 15BC

[a] Life cycle = 20 yr. discount rate = 0.06, interest rate = 0.085, downpayment = 20%, rate of fuel increase = 0.01.

Other Case Studies

The case study of Colorado Springs illustrated the complexity in predicting the optimal cost-effective building configuration under various tax-credit scenarios. Under rational conditions, one would adopt a particular building design which accounts for the stream of changes in variables such as rates and tax credits over the life of the solar application or building. The results for Colorado Springs depended on the utility area's climate, the utility characteristics, and many building design and cost parameters. The individual effect of each of the variables from a national perspective has not been determined clearly. It is unclear what sensitivity exists in the use of other locations, utilities, or building costs. The next subsection examines the sensitivity of these variables in order to illustrate the particularity of the Colorado Springs case study.

The New England Electric System in Worcester, Massachusetts. The first variation in utility location is the use of another winter-peaking utility, the New England Electric System, (NEES) of central Massachusetts and Rhode Island. The study point—Worcester, Massachusetts— represents the largest city within the service area. Unlike the Colorado Springs Department of Public Utilities, NEES is a privately owned utility. The other major difference between the two utility companies is the larger dependence on oil as a fuel source in the NEES system.

The differences in degree-days between Colorado Springs and Worcester is exhibited by a 10 percent higher consumption of energy in the base building in Worcester than in the base building in Colorado Springs. The higher insolation levels in Colorado Springs are also reflected in the thermal results; the same 35-m^2 collector collects about 2,000 kWh more useful energy in Colorado Springs than it does in Worcester.

The financial results for four buildings in New England were compiled, as was done in Colorado Springs, and are shown in Table 5–12. The four buildings include:

1. The base building with shade, 2 × 4 in insulated stud wall, 6-in attic insulation, light room construction, and 16 percent south-facing double-glazed windows
2. The Energy-conservation building, with 2 × 6 in insulated stud wall, 12-in attic insulation, no total wall, light room construction, overhanging shade and 16 percent south-facing double-glazed windows
3. The masonry building with 8-in concrete block, 4-in total wall, heavy room construction, 12-in insulation, no overhanging shade, winter/summer variable thermal shutter, and 80 percent double-glazed windows
4. The minimally insulated building, with 2 × 4 in uninsulated stud wall, 6-in attic insulation, light room construction, overhanging shade, 16 percent south-facing double-glazed windows

The most significant difference between the results of Colorado Springs and Worcester is that even in the first year in Worcester, the investment in additional insulation beyond the base building is justified because both the all-electric energy-conservation building and the all-electric masonry building have lower first-year costs than the base building (table 5–13). Any of the four tax credits previously examined further emphasize the value of insulation, with the California-style tax policies creating a clear advantage for the masonry building. Although the lowest-cost alternative is the masonry building all-electric version, investment in a solar hot-water system is also cost-effective in comparison to the base building all-electric version.

The real effect of higher electric rates for Worcester is seen in the life-cycle analysis (table 5–14) where the price of fuel becomes so great that

Table 5-12
Consumer Cost-Effectiveness for New England Electric System

	Load Factor	First-Year Costs to Consumer ($)		Consumer Life Cycle Cost ($)					
				Low Future Scenario[b]		Medium Future Scenario[c]		High Future Scenario[d]	
		Without Tax Credit	With 55% Tax Credit[a]	Without Tax Credit	With 55% Tax Credit[a]	Without Tax Credit	With 55% Tax Credit[a]	Without Tax Credit	With 55% Tax Credit[a]
Base building	0.17	1,357	1,012	14,700	11,713	17,193	13,904	22,222	18,538
	0.33	1,122	925	14,481	12,766	17,888	15,992	26,602	24,479
	0.37	966	928	13,559	13,207	17,158	16,770	26,908	26,489
Energy-conservation building	0.15	1,349	1,006	14,265	11,290	16,424	13,135	20,513	16,803
	0.33	1,086	847	13,573	11,456	16,608	14,267	24,194	21,573
	0.38	929	849	12,664	11,890	15,878	15,045	24,484	23,551
Masonry building with winter-summer shutters	0.08	1,531	1,191	15,299	12,324	17,227	13,938	20,192	16,509
	0.24	1,116	783	12,420	9,445	14,624	11,335	19,452	15,769
	0.31	959	700	11,490	9,108	13,895	11,261	19,742	16,792
Minimally insulated building	0.26	1,960	1,613	23,645	20,670	28,562	25,273	40,344	36,660
	0.36	1,827	1,653	24,833	23,326	23,341	29,475	47,824	45,958
	0.38	1,671	1,655	23,903	23,760	30,412	30,253	48,114	47,936

[a] Tax credit extended to both energy conservation and solar.
[b] Low future scenario: Discount rate = 10%, electricity escalation = 6%
[c] Medium future scenario: Discount rate = 8%, electricity escalation = 7%
[d] High future scenario: Discount rate = 6%, electricity escalation = 10%

Table 5–13
**Impact of Tax Credits on Consumer Cost-Effectiveness in the New England
Electric System; First-Year Cost Analysis**

Tax Policy	Lowest Cost Alternative
1. No tax policy	All-electric energy-conservation building
2. Federal solar tax policy (originally proposed) (Solar, 30% of first $2,000 and 20% of next $8,000; Conservation, 20% of first $2,000)	All-electric energy-conservation building
3. Federal solar = energy conservation tax policy (*hypothetical*) (Solar and energy conservation, 30% of first $2,000, 20% of next $10,000)	All-electric energy-conservation building
4. California solar tax policy (*existing*) (55% of active and passive solar and energy conservation in conjunction with solar up to $3,000)	All-electric masonry building
5. California energy conservation = solar tax policy (*hypothetical*) (same as 4 but includes all energy conservation)	All-electric masonry building

All buildings have overhanging shade and/or thermal shutters.

Table 5–14
**Impact of Tax Credits on Consumer Cost-Effectiveness in the New England
Electric System; Life-Cycle Cost Analysis, Lowest Cost Alternative**

Tax Policy	Low Future Scenario	Medium Future Scenario	High Future Scenario
No tax policy	All-electric masonry building 14C	All-electric masonry building 14C	Solar hot-water masonry building 14B
Californias solar tax policy extended to include all energy conservation and solar, 55% up to $3,000	All-electric masonry building 14C	All-electric masonry building 14C	Solar hot-water masonry building 14B

Low future scenario: Discount rate = 10%, electricity escalation = 6%
Medium future scenario: Discount rate = 8%, electricity escalation = 7%
High future scenario: Discount rate = 6%, electricity escalation = 10%

nearly any investment which lowers energy consumption has a lower cost
than the base building. The best investment of the ones examined remains
the all-electric masonry building in all but one case. In the high future
scenario, costs of fuels are inflated so high that investment of a solar hot-
water system on the masonry building is justified.

Utility Impact

The performance of the different buildings at the time of utility peak in Worcester is slightly better than in Colorado Springs. The weather conditions at time of peak in Worcester are even more related to extreme cold than was the case in Colorado Springs. Performance of the solar systems at utility peak is not as poor as in Colorado Springs, yet the load factors of the solar-heated buildings are still less than half those of the identical all-electric versions. As in Colorado, investments in energy conservation and solar hot water do not significantly lower load factors. As a consequence of low load factors for solar-heated buildings, there is a commensurate imbalance in utility net revenues.

Societal Impact

The societal benefits of investment in energy conservation are realized in the first year despite all ignorance of incurred social costs and future fuel-price rises. Masonry buildings also have greater benefits to society.

The benefits to society are more clearly shown in the life-cycle cost assessment in table 5-15. The clearest result is the poor economic performance of the minimally insulated building. This building, which represents most present housing stock, had a societal cost twice that of any other building and utilized over 300,000 kWh and 4.6 kW of additional electricity and capacity in twenty years. The savings to society from investment in wall insulation at a cost of $344 (in a new house) is $8,357 in present worth. There is a ratio of nearly 20:1 of present-worth benefits to present-worth costs ($8,357:422). In addition, savings also are realized by society in the large reduction of kilowatt hour consumption and kilowatt capacity requirements. Large benefits also are realized by society for further investment in energy conservation in 2 × 6 in insulated stud wall, 12-in attic insulation, and high-mass construction and passive solar as in the masonry building. Benefits to society of investment in either a 5.6-m² solar hot-water heater or a 35-m² solar heating and hot-water system do not exceed the benefits from the energy-conservation investment.

Public Service Company of New Mexico at Albuquerque

The case of Public Service Company of New Mexico (PSCNM) is an example of a summer-peaking utility. In this utility the primary concerns are to reduce summer air-conditioning and hot-water loads. Active solar heating, therefore, does not reduce peak demands but does reduce annual sales. This lowers the utility-system load factor and utility operating efficiency.

The PSCNM service area has a large number of heating degree-days and has experienced several system peaks during the winter months.

A trend in building cost-effectiveness similar to those found elsewhere emerges in Albuquerque (table 5-16). The all-electric energy-conservation building has the lowest first-year cost. The all-electric masonry building has the lowest first-year cost when California tax policies are applied. The significant change between Albuquerque and other locations is that under life-cycle analysis with high future scenario, the solar-heated base building is the lowest cost alternative. This building and the solar-heated energy-conservation building exceed the performance of the all-electric building.

One reason for this outcome is the large amounts of cooling required by the masonry building. Closer examination of this cooling load reveals that the heavy thermal mass combined with cool summer nights effectively eliminates almost all summer cooling requirements. The inclusion of 80 percent glass, however, causes significant winter overheating problems. Since no provision is made in the model to ventilate rather than air-condition, the masonry building cooling load remains. The example hampers evaluation of true cost-effectiveness of building, but does illustrate a critical problem of passive buildings—overheating. It is possible that 40 percent south-facing glazing will eliminate much of this overheating without significantly increasing heating loads. This case was not examined, however.

A negative utility impact appears to exist for solar heating in Albuquerque, particularly as regards the minimal contribution of the solar heating system to summer peak-load reduction. The opposite appears to be true of solar hot-water systems, which improve the load factor in several cases.

Societal considerations appear again to favor initial investment in energy conservation by the consumer (table 5-17). Societal benefits for active and passive solar do however exist. One particular benefit of a solar hot-water heater is the benefit of the solar system in reducing peak electric loads. Although the savings in kilowatt hours for the passive energy investment (row 3) is greater than that of the solar (row 4), the dollar savings to society per dollar investment is greater for the solar water heater because virtually all the reduction is in on-peak electricity.

Pacific Gas and Electric

The first-year analysis shows that no building can presently compete with electricity in a building that very closely conforms to new standards in California called a *new-standards* building. The fact is that even with the existing solar tax credit, or a hypothetical tax credit which extends coverage to energy-conservation investment not in conjunction with solar, no

Table 5-15
Comparison of Societal Benefits and Costs among Various Investments in Technology; New England Electric System

Investment	Kilowatts Saved between Alternatives	Kilowatt Hours saved between Alternatives (1,000 kWh)	Savings in Life-Cycle Cost in Dollars of Utility Generation Capacity Costs	Savings in Life-Cycle Cost in Dollars of Utility Variable Costs	Total Life-Cycle Savings to Utility ($)	Life-Cycle Cost to Consumer of Investment (Original Investment Cost)	Ratio of Savings to Cost
Insulation of uninsulated 2 × 4 in wall (building 4 versus building 1)	4.66	324	1,926	6,331	8,357	422 (344)	19.6:1
Insulation to 2 × 6 in wall stud and 6-in additional attic insulation (building 1 versus building 2)	0.83	50	343	1,609	1,952	800 (651)	2.44:1
Masonry, 80% glass and thermal shutters (building 1 versus building 3)	2.08	178	860	5,639	6,499	3,750 (3,054)	1.73:1
Solar hot-water system versus conventional domestic hot water	0.28	35	115	1,544	1,569	3,069 (2,500)	0.54:1
35-m² collector, 2.65-m³ storage on base building versus conventional electric building	1.22	248	504	8,256	8,760	10,251 (8,268)	0.85:1

Table 5-16
Consumer Cost-Effectiveness for Albuquerque

	Load Factor	First-Year Costs to Consumer ($)		Consumer Life-Cycle Costs ($)			
				Low Future Scenario[b]		High Future Scenario[c]	
		Without Tax Credit	With 55% Tax Credit[a]	Without Tax Credit	With 55% Tax Credit[a]	Without Tax Credit	With 55% Tax Credit[a]
Base building	0.13	1,098	753	12,594	9,305	14,909	11,226
	0.68	882	690	13,558	11,717	19,780	17,717
	0.70	690	658	12,183	11,849	19,044	18,670
Energy-conservation building	0.11	1,142	799	12,855	9,566	14,991	11,429
	0.65	870	641	12,792	10,558	18,232	15,730
	0.68	678	608	11,417	10,690	17,496	16,681
Masonry building with winter-summer shutters	0.72	1,481	1,140	17,023	13,733	20,305	16,622
	0.86	949	614	12,204	8,914	16,001	12,317
	0.80	757	542	10,828	8,652	15,265	12,829
Minimally insulated building	0.35	1,334	987	17,199	13,914	22,392	18,708
	0.91	1,356	1,185	22,542	20,908	34,209	32,379
	0.88	1,164	1,152	21,166	21,040	33,473	33,331

[a] Tax credit extended to both energy conservation and solar.
[b] Low future scenario: Discount rate = 10%, electricity escalation = 6%
[c] High future scenario: Discount rate = 6%, electricity escalation = 10%

Table 5-17
Comparison of Societal Benefits and Costs among Various Investments in Technology, Albuquerque

Investment	Kilowatts Saved between Alternatives	Kilowatt Hours Saved between Alternatives (1,000 kWh)	Savings in Life-Cycle Cost in Dollars of Utility Generation Capacity Costs	Savings in Life-Cycle Cost in Dollars of Utility Variable Costs	Total Life-Cycle Savings to Utility ($)	Life-Cycle Cost to Consumer of Investment (Original Investment Cost)		Ratio of Savings to Cost
Insulation of uninsulated 2 × 4 wall (in building 4 versus building 1)	1.04	237	431	6,196	6,227	422	(344)	14.8:1
		(Savings per $10,000 investment = $192,645)						
Insulation to 2 × 6 in stud wall and 6-in additional attic insulation building 1 versus building 2)	0.27	37	110	950	1,060	800	(651)	1.3:1
		(Savings per $10,000 investment = $16,283)						
Masonry, 80 percent glass and thermal shutters (building 1 versus building 3)	1.20	123	495	1,478	1,973	3,750	(3,054)	0.5:1
		(Savings per $10,000 investment = $6,450)						
Solar hot-water system versus conventional domestic hot water	0.64	84	263	2,665	2,928	3,070	(2,500)	1.0:1
		(Savings per $10,000 investment = $11,712)						
35-m² collector, 2.65-m³ storage on base building versus conventional electric building	0.73	255	302	7,391	7,693	10,152	(8,268)	0.8:1
		(Savings per $10,000 investment = $9,305)						

building has lower first-year cost than the $510 figure for base building C. The three buildings that do come closest are base building A; 2 × 6, no total wall, C; and 2 × 4 with total wall, C. All are within $40 of the base building C. However, the two 2 × 6 buildings are not competitive under the existing tax-credit policy.

The cost to the utility and the total costs for each of these buildings present different results.

The tax credit granted to solar consumers represents a loss in revenue for the state of California, and therefore the total costs must include this loss of revenue. On the other hand, the total cost does not account for societal benefits which accrue from the use of energy conservation and solar energy. (These benefits include less dependence on foreign energy sources, less depletion of scarce resources, improved balance of trade, and less environmental degradation.) To the extent that these benefits are not reflected in the market price of energy, the total costs are biased against energy conservation and solar investment. This unaccounted contribution to society forms the economic rationale for a tax credit being implemented.

Examination of the total costs for each of the buildings reveals again that base building C is the least-cost alternative. From the total cost picture, investment in energy conservation has more immediate benefits than investment in solar energy. This is demonstrated by the fact that the next least-cost alternatives are the two 2 × 6 in buildings with all-electric heating and hot water.

The most significant evidence of total saving is shown when the old residential building stock is retrofitted to the new California building standards. (The base building simulated here has incorporated new California building standards plus some minor energy-conservation additions.) For example, the total cost for the all-electric uninsulated-wall building drops by 36 percent when it is retrofitted to the new standards. In contrast to this, total costs increase 17 percent when investment in a solar hot-water system is made.

There appears to be very little difference between the existing rate and an experimental Time-of-Use (TOU) schedule on the relative cost-effectiveness of buildings. In general, the utility is absorbing more of the cost of service of space conditioning and domestic hot water under the time-of-use tariff. One significant finding is that the total cost to the consumer is significantly reduced for the existing housing stock under the TOU tariff. It can be demonstrated, by comparison of the total cost to the existing housing stock owner under the two tariffs, that the TOU tariff may postpone investment in solar and/or energy conservation. An additional consideration is that a general residential TOU tariff may make thermal energy storage a lower-cost investment than other alternatives. For a utility such as PG&E, which in the future may be more constrained by energy availability than generating-capacity availability, the commercialization of thermal energy storage may have significant impacts.

The analysis assumes no cross-elasticity of demand between peak and off-peak use. However, solar-with-storage (or storage alone) consumers may be more inclined to invest in solar or thermal energy storage systems under TOU pricing since they will be able to take advantage of off-peak rates. These general conclusions are not novel, but they are quantitatively evident, as shown by the figures in the above tables.

Under the existing residential rate schedule, the least-cost alternative to the consumer with the tax credit is base building A, the new-standards building with a combined solar hot-water and heating system. The reduction in life-cycle costs by investing in the combined solar equipment as compared to the all-electric system is $1,836. It is interesting to note that life-cycle cost to the consumer is approximately the same for base buildings B and C, and both energy-conservation (2 × 6) buildings A, B, and C. The differences evident for the existing rate, between base building A and these other buildings, are not as pronounced under the experimental TOU rate.

An interesting result is the worse performance of the uninsulated-wall building under a life-cycle cost analysis. The cost to the consumer of the uninsulated building is almost twice the amount for the base building. The greatest benefits are realized from retrofitting the home for insulation rather than from solar energy investment.

One important point gleaned from these tables is the potential savings achieved by passive solar energy investment. The masonry building with 40 percent glass and shutters has a low total cost. Only base building C has a lower cost. With the tax credit, the solar passive building is more cost-effective than every other building except the combined solar hot-water and heating system on base building A.

Los Angeles Department of Water and Power (LADWP)

First-year cost analysis and life-cycle analysis were performed for the various technical configurations with the assumption of three different LADWP rate schedules. In the absence of a tax credit, the all-electric base building has by far the lowest first-year cost. With the existing tax credit, base building A and base building B become more competitive, although the all-electric base building is still the most cost-effective. Should the tax credit be extended to energy-conservation measures not in conjunction with solar, then the energy-conservation building (2 × 6 in walls with total wall and shade) has a first-year cost lower than that of the base buildings with solar investment.

The first-year cost analysis under the enjoined flat-rate schedule shows that since the unit price per kilowatt-hour is higher under this schedule, the first-year costs to the consumer are higher in all cases. Under this rate, investment in energy conservation and solar becomes more competitive with the all-electric base building used for comparison.

In each of the four building structures, compared to the existing rate, the TOU rate lowered the all-electric building (C) first-year costs but had little impact on the solar heating and hot-water building (A) costs. Wide implementation of this tariff would result in an investment in energy conservation being favored over investment in solar energy, the tax credit notwithstanding.

Assumptions used for the life-cycle cost analysis are those of the medium future scenario. With the tax credit, base building A is the least-cost alternative, as was the case for PG&E. Maximum benefits for the state accrue when investment is made in energy conservation to bring the existing housing stock up to the level of the new standards. Total life-cycle costs decrease by $5,666, or from $14,393 to $8,731, when minimal investment is made to bring the all-electric uninsulated building up to the new standards. This huge savings obtains from only a $371 capital investment. On the other hand, a $5,652 investment in a solar hot-water and heating system for the uninsulated building actually results in a $944 increase in total life-cycle costs. A possible perverse effect of the existing tax-credit policy may be to encourage the existing housing stock owner to "insolate" his home rather than "insulate" it!

The masonry building as modeled for LADWP does not appear to be very cost-effective. Improvements to the thermal performance of this building suggested previously in the PG&E case study no doubt would make this alternative more cost-competitive. These improvements include increasing south-facing glazing and thermal shutters.

Implications for Public Policy

The continued drain of U.S. dollars from the purchase of foreign oil requires effective policy to reduce fossil-fuel consumption and promote nonpolluting energy sources. The benefits of solar energy to this end are significant, and accordingly federal policy should view favorably solar energy's rapid development. As evidence one can cite the recent passage of a national energy plan which included tax credits for solar energy as a way of promoting the rapid commercialization of solar energy and overcoming the significant barriers to its development.

Tax credits for solar energy in part compensate for the present subsidies to existing energy technologies. The tax credits also acknowledge the benefits received by society that exist external to the marketplace. The incentives overcome part of the resistance to the large initial investment that solar energy requires. Yet several issues arise with regard to the tax credits, particularly concerning the emphasis on active solar energy rather than energy conservation and passive solar energy design.

While both solar heating and energy conservation play an essential role in meeting future national energy requirements, the case study for Colorado Springs, further qualified by other utility analyses by these authors, has shown that additional amounts of energy conservation beyond present practices provide greater societal benefits, at lower costs. These greater benefits can be seen by the greater savings in dollars, energy, and generation capacity from energy conservation as compared to active solar heating. While other impacts and benefits beside these three factors exist, there does not appear to be, outside of particular idiosyncrasies, any significantly greater benefits from solar heating than from energy conservation and passive solar energy design.

In contrast, the incentives provided by the national tax-credit policy are far greater for active solar energy ($2,200 maximum) than for energy conservation ($300 maximum). The impact of this situation may be the growth of solar energy at the expense of energy conservation.

The lack of equivalent tax benefits for active solar heating and energy conservation stems from the belief that energy conservation does not require as much promotion as solar heating. In addition, a bias against energy conservation exists which downplays energy conservation as an important aspect of energy policy. Despite the fact that a unit of energy saved is at least equivalent to a new unit found, federal commitment to energy conservation may be insufficient. Energy-conservation practices in many homes are well below the levels found economical in this book and societal benefits could justify further promotion.

In Colorado Springs, greater tax credits for active solar do promote active systems at the expense of energy conservation. Even where this advantage is insufficient to make active solar energy the least-cost alternative, the perceived tax savings of active solar may be sufficient to dissuade energy-conservation investment. Perhaps what is most disturbing about the present federal tax-credit policy (California's tax policy is an improvement in this respect) is the absence of tax credits entirely for passive solar energy which was found to be an extremely effective energy-reduction technique.

The problem with tax credits is that they must contain very specific language. Passive techniques do not fit into easily definable categories and are thus ignored. Another difficulty is that tax credits promote active solar energy without regard for additional problems active solar may cause to electric utilities.

Utility Financing of Solar Applications and Energy Conservation[a]

Among the several generic incentive programs proposed to accelerate the widespread use of decentralized solar energy is a class of financing alternatives administered by regulated public utilities. It is the purpose of this section to analyze various forms of utility solar finance. This analysis delineates the complexity of the regulatory issues involved with any scheme which uses public utilities as financial intermediaries for decentralized solar.[7]

At this point we discuss various types of utility solar financing arrangements. The focus is on the costs of each arrangement to the utility. The discussion then broadens to consider the generic impacts of utility solar financing on both the customer and the corporation.

Utility solar financing is an intrinsically complex issue. This stems from the central position already occupied by public utilities in the existing energy distribution and marketing system. Whether utilities finance solar energy or not, they are impacted intermediaries in *any* plan to accelerate the commercialization of on-site systems. If utilities are external to the solar incentive process, then they may well emerge as a constraint on the solar engineering economic optimization. This constraint appears in the form of utility pricing policies for backup energy. The preceding chapters have shown the effect of utility prices on solar design choices. In turn, the utility pricing of backup energy is determined by the changes in load shape and cost of service imposed on the utility by widespread adoption of solar technology. It is not at all clear, however, that the real world will react as flexibly as predicted by an optimization model in which all inputs are known with certainty.

Utility participation in solar energy financing is likely to create a different set of adjustments to the energy-supply planning process than if there were no such participation. These differences would be due to both the potential scale of a solar program as a whole and the optimal design of individual systems. Current incentives for solar energy in the form of tax credits have had a highly limited effect. In California, for example, the state income tax credit has produced a response which is highly skewed toward upper income groups. Table 5-18 shows recent data on this trend. It is clear from this table that 75 percent of all applicants had adjusted gross incomes of over $20,000 per year, and 30 percent were over $40,000 per year. In certain markets, such as the multifamily rental markets, there is no incentive, even with potential tax credits, to invest in solar energy application. Bezdek, Hirshberg, and Babcock (1979) attribute the lack of incentive in the apartment sector to the investment goals of owners and the structure of the

---

[a] This section was written by Edward Kahn.

Table 5-18
California Solar Energy Tax-Credit Applications by Income Level

Adjusted Gross Income ($)	Approximate Number of Solar Credit Applications	
Less than 9,999	885	
10,000–14,999	1,155	4,150
15,000–19,999	2,110	
20,000–24,999	2,855	
25,000–29,999	2,902	
30,000–39,999	4,504	16,382
40,000 and over	6,061	

California Franchise Tax Board Data cited in Testimony of T.A. Keefe, 1979.

tax code. A solar financing program administered by regulated utilities would, in principle, be addressed to larger markets than the tax-credit approach currently attracts.

A more subtle but equally important effect of utility solar financing is the potential for more optimal system design. A private or corporate decision to invest in solar energy will be based on current energy prices. Insofar as these prices are public-utility rates, the investment decision and the optimum economic design of a solar system will be biased away from solar systems which displace large amounts of energy. The reason is simply that current public-utility rates are based on average historical costs. These are lower than the marginal cost of new supply. Economic efficiency is achieved by trading off options at marginal cost. Under current market arrangements, there is no actor who can ensure that the appropriate value of displaced energy will be reflected in the private decision process. Utilities, however, are in a position to compare marginal costs. Other things being equal, this would result in more efficient and presumably larger-scale investment in solar systems.

Although utility solar financing might well accelerate the adoption of on-site solar systems, there are risks and costs associated with such arrangements. The main risk is the potential for monopolization associated with the scale of such activity. This risk has been characterized in a variety of ways in chapter 6 by R. Noll (see also Laitos and Feuerstein, 1979; Schmalensee 1978). Apart from a generalized antipathy toward monopoly, there are real and potential costs of using the utilities as a principal financial intermediary for solar commercialization. In principle, banks and other conventional financial institutions have a lower cost of money than regulated utilities. This is readily apparent by a comparison of capital structures. Banks are capitalized at roughly 95 percent debt (that is, deposits) and only 5 percent equity. Utilities typically have about 35 percent equity capital which is costlier than debt or preferred stock. The cost difference between debt and equity is at least several percentage points. By encouraging utility solar financing, society would therefore be choosing an intrinsically more expensive source of finance than might be available through conventional means. This extra social cost must be weighed against the potentially greater market available to on-site solar through utility finance.

Other less readily quantifiable risks of utility solar financing include the potential for economic distortions induced by a tendency toward excess capitalization or by possible motives to cross-subsidize solar activity from other investments. The risk of subsidization of solar occurs where the utility underprices rather than overprices. Assessing the importance of this risk requires an analysis of the vendor market for on-site solar. If such analysis indicates that utility financing might involve unfair subsidies, regulatory alternatives exist to limit this danger.

In the analysis which follows, various generic arrangements for utility solar financing are surveyed. Each alternative is characterized by its main advantages and disadvantages with regard to impact on customers, the utility, and more general social concerns. After the generic alternatives are discussed, individual issues are addressed. These include (1) analysis of economic impact on utilities of involvement in solar energy financing and (2) economic impact on nonparticipating utility customers of solar energy financing programs.

Impact of Utility Financing on Demand

In the previous section we considered the desirability of public-utility financing schemes from a utility point of view and concluded that such proposals would not be disadvantageous as an investment. In this section we consider the extent to which such proposals will actually stimulate demand for solar installations.

To see clearly the workability of public-utility financing proposals, we first consider the record of private lending institutions offering special terms to borrowers seeking to install solar systems on their homes. A number of private lending institutions in California offer programs in which consumers can borrow at reasonable rates for solar installations. One such institution, San Diego Federal Savings and Loan, offers the prime residential interest rate for solar installations. For homeowners who have their first mortgage with San Diego Federal, they will extend the term of the existing first trust deed with the result that there is no increase in the current monthly rate paid by the borrower. All the lending institutions which have solar programs, including some which provide loans at one percentage point below the normal home-improvement loan, have advertised and promoted their program. In one survey of those institutions offering special programs, all eight institutions term their programs a failure.[8] Although the demand for loans for other home improvements is good in those institutions, the demand for solar loans, at lower rates, is nonexistent.

The reasons for the failure of private loan programs to stimulate the demand for solar improvement loans is undoubtedly related to problems in demand experienced by the solar industry at large, and this is beyond the scope of this discussion. The record of the private programs is, however, important in determining the potential effectiveness of public-utility financing proposals. It is clear that utilities do have characteristics which may make them more appealing to consumers seeking funds for solar systems. Public utilities could offer more "complete" services, including advice on contractors, system choice, maintenance services, and convenient financing services. All these factors could stimulate consumer demand. Utilities certainly could be persuaded to undertake these additional tasks if they were compensated with rate-base treatment. In a proposal by PG&E in California before the California Public Utilities Commission to OII 42, the company suggested that it could aid the owners of multifamily dwellings in determining the cost-effectiveness of solar systems, choosing a solar system design, selecting a contractor, and choosing an institution from which to obtain a loan. In terms of regulatory treatment, PG&E suggested that "all program-costs—including administrative expenses, provision for bad debts, and promotion—not covered by solar customer payments will be expensed for rate making purposes" (Pacific Gas and Electric Co., OII 42, p. 8). It appears, therefore, that utilities could provide services beyond what lending institutions are capable of giving. With aggressive advertising and good marketing techniques, public utilities appear to be in a better position to stimulate demand for on-site solar installations.

A second major area of consideration in determining the advantages of utility financing is the rate of interest which will be charged consumers and the payback period allowed. It is clear that the consumer demand for solar loans is inversely related to the rate of interest on the loan. Although the precise price elasticity of demand for solar loans has not been determined, we would expect variations in demand if the rate charged by utilities differed significantly from the going rate obtained from private institutions. Several studies have been done to determine the workable rate that should be levied by the utilities, most notably Kahn and Schutz (1978) and Czahar (1979).

Kahn and Schutz (1978) determine the appropriate finance charge by considering the rate on a risk-free investment plus the market price of risk multiplied by the volatility of solar investments (beta). Kahn and Schutz estimate beta for on-site solar installations to be approximately 0.15. If we use an estimate of 8.8 percent as the market price of risk, the appropriate finance charge should be 1.4 percent above the "riskless" long-term Treasury Bill rate. Kahn and Schutz therefore estimate a finance charge that is substantially below the going rate for solar loans from private institutions. (This is discussed in somewhat more detail below.) If this rate is accurate, we would expect a significant increase in the demand for solar loans if the price elasticity of demand for the loans were sufficiently elastic. It must be noted, however, that the finance charge determined by this method does not consider what return the utility will earn by making the loan. This depends on the constraints set by the regulatory commission.

A second problem with the Kahn and Schutz estimate is whether the market will actually evaluate the risk as indicated by the Capital-Asset-Per-centage-Model (CAPM) approach. Thus the ability of the utility to use too low a finance charge assumes that it can obtain specific funds through debt financing at the appropriate rate and that funds obtained at higher rates for more risky investments will not affect the finance charge for solar users. It becomes clear that the regulatory commission plays an important role in actually determining the finance charge. Decisions made by the regulatory commissions, of whether to subsidize the loan program or allow the utility to extract the costs from the consumers, will certainly have an impact on the loan programs.

If we assume that the utility will be allowed to treat the solar investment as any other, we can determine a finance charge that is consistent with the financing of conventional plant. Czahar (1979) has done a study to determine the appropriate rate that the utilities should charge under this assumption. Czahar obtains his estimate by considering the total cost of capital to the utility. This is done simply by summing the weighted cost of the three sources of financing available to the utility plus the impact of state and federal taxes, with the total discounted for the utilization of the investment tax credit (assuming 60 percent utilization of the credit). Looking at the entire capital pool, Czahar estimates that the pretax cost of capital—the effective rate that would be charged consumers—is approximately 13.52 percent. This estimate is consistent with the finance charge levied by private lending institutions for solar loans, and we may expect similar success of the program.

It should be noted, however, that the utility financing program would be able to provide, in addition to the services discussed above, long-term loans to consumers that may effectively reduce the monthly charge to an affordable one. Czahar points out that a $2,000 loan for 20 to 30 yrs at 13 percent would require monthly payments of only $22. Such services will undoubtedly contribute to the success of the program. It should also be noted that Czahar's estimates of the appropriate finance charge are themselves open to qualification. First, Czahar does not consider the fact that the cost of capital for solar investments will be less than the cost for conventional systems. Based on the findings of Willey (1978) and Kahn and Schutz (1978), we have seen that utilities should be able to obtain debt financing at more reasonable rates because of their solar investments. It would also be important to determine if the sources of capital would charge in proportion to total financing as the utility invested in solar technologies. In his model of alternative energy systems Willey assumes that the ratio of debt to total finance will not change with the inclusion of solar systems, but that the ratio of internal to total finance will be more favorable when solar investments are included. The changes in these ratios should change the cost of the capital pool from which Czahar makes his estimates.

The second point of qualification about Czahar's estimates is that he allows the utilities to earn the average return on their investment, unlike Kahn and Schutz. Czahar's estimate of the cost of debt, preferred and common, is based on the return which the utility must obtain in order to pay dividends and earn a positive rate of return. If the cost is not borne by the solar consumers, because of regulatory manipulation, the cost of the solar loans would be substantially less.

We can see, therefore, that the cost of the loans to solar consumers can vary substantially depending on the specific financing arrangements in the proposal. These financing arrangements are a matter of policy choice and are considered briefly in the following section.

Specification of Generic Utility Solar Financing Alternatives

Traditionally, financing has offered opportunities for innovative arrangements that are limited only by uncertainties associated with the legal status of the proposed instrument. Thus, many variations of basic alternatives are possible for any financing mechanism. The current investigation by the California Public Utilities Commission (CPUC) into utility solar financing has produced a catalog of fourteen variations (Testimony of T.A. Keefe 1979). It is doubtful if this exhausts the range of permutations and combinations of specific program features. Rather than enumerate all possibilities, it will be convenient to catalog the major classes of alternatives and their main features. In any particular situation, conditions will favor some combination of the main features.

We begin by contrasting the role of solar capitalization by the utility with the role of financing. This is followed by an analysis of the Pacific Power and Light Company's residential energy efficiency rider. This plan is thus far the most far-reaching, private utility sponsored end-use efficiency program in the nation and has enjoyed widespread acceptance by customers. It has been proposed as a model for other utilities, and therefore it deserves special attention. We conclude by examining the role of leasing arrangements, the case for creating special utility solar subsidiaries, and the role of special bonding authorities.

Capitalization versus Financing. The standard accounting treatment of any utility capital investment is that all appropriate costs for materials and labor are added to the undepreciated rate base to earn the allowed rate of return on capital. In the special case of utility investment in on-site solar, such treatment may or may not involve cross-subsidization of solar investments by nonsolar users. By "rolling in" all solar costs into a rate base common to all customers, the nonsolar user pays an incremental cost for conventional service over and above what he would have paid without the utility solar investment. If this increment is greater than the marginal cost of new conventional supply suitably allocated, then solar users may be said to be subsidized. The regulatory remedy for this situation is straightforward. Solar investments can be capitalized in a separate account charged to solar users only. While this would avoid cross-subsidization, it has the consequence of charging marginal costs for solar energy, but only average costs for conventional supply. Although public-utility rates should not involve cross-subsidies in theory (Bonbright 1961), in practice it goes on to a considerable extent. The main practical concern in this regard is the magnitude of such subsidies. For typical conditions in the gas industry, it has been shown that "rolling in" will have a small impact on nonsolar rates (Boyd et al. 1978). The relatively small fraction of utility capital that would be devoted to solar is the reason for this result. Under widespread implementation, this effect could be considerable. A more stringent criterion concerning cross-subsidization is discussed in connection with the Pacific Power and Light Company plan.

A potential complication of any capitalization approach is the risk of "gold plating." It is possible that utility ownership of solar would be biased toward expensive, overdesigned systems that are excessively capital-intensive. This is really just an instance of the Averch-Johnson thesis (Averch and Johnson 1962) that rate-of-return regulation induces a bias toward capital.

In terms of the engineering/economics of active solar space heating, for example, gold plating might take the form of underinvestment in glazing, insulation, weather stripping, and so on. As we have already shown, such conservation investments reduce the thermal load that must be supplied by collectors and are considerably less expensive (see also, Booz, Allen and Hamilton, Inc. 1979). Unfortunately, the utility might have difficulty qualifying for federal tax benefits such as rapid amortization and investment credit from residential conservation investment. The Internal Revenue Service (IRS) grants such benefits only to investment that is made on the owner's property and dedicated to his use (Amaroli, Personal communication). A utility's residential meter passes these tests, but conservation and solar investments would have more difficulty. Without equal tax treatment for all energy options, the utility could not be expected to make efficient choices.

The generic alternative to capitalization is a strictly financial role for the utility. In this role, the utility would act as a bank which makes loans for a predetermined period at a fixed rate of interest. The costs of such a program depend critically on the choice of loan period and interest rate. The appropriate loan period should be the economic lifetime of the solar system. Unfortunately, there is considerable uncertainty about this period. Choice of a relatively long lifetime, say 20 years, would almost certainly mean that individual components would require earlier replacement. In table 5-19 estimated component lifetimes for solar hot-water heating systems are listed based on CPUC recommendations (California PUC Energy Conservation Team 1977). Some provision for the cost of replacing components must be made if the financing is based on a 20-year lifetime.

Economic lifetime is also an important parameter under capitalization. In that case, it represents the length of time for which the capital investment is in the utility rate base. Lifetime also determines the depreciation schedule with any method of depreciation. Depreciation will reduce the rate base under capitalization, but is irrelevant to the utility under financing.

Fixing an appropriate interest rate under financing can be approached in a number of ways. The standard procedure in the conventional economic analysis of utility investment projects is to calculate a fixed-charge rate to be charged annually against capital cost to yield the pretax, weighted-average cost of capital (EPRI 1978a). Thus, a capital structure is assumed, the cost of each kind of capital is estimated, and tax effects are added (see table 5-20). Fixed-charge rates will vary across regulatory jurisdictions depending on the treatment of federal tax preference. Utility commissions which require "flow-through" of investment tax credit and accelerated deprecia-

Table 5-19
Solar Hot-Water Heating System Component Lifetimes

Item	Estimated Lifetime (yr)
Solar collectors—copper type	20
Pumps	10–15
Valves	5
Solar hot-water storage tank	20
Backup hot-water heating system	10
Controller	10
Associated copper plumbing	20

tion to customers will tend to see lower fixed-charge rates than commissions where tax preference is captured by the utility. This subject is discussed in some detail below. For now it is sufficient to observe that for solar energy financing, assuming that no tax credits would be available to the utility, the pretax cost of capital is currently in the range of 17 to 22 percent. The conventional interest rate so determined would be the same under capitalization or financing.

Because the pretax cost of capital is so high compared to bank rates, utility solar financing would not be particularly attractive. However, it is not at all clear that the standard procedure used to determine fixed-charge rates, as sketched above, is the appropriate tool for utility economic analysis. Public utilities may be thought of as a portfolio of investments, the sum total of which provides a service to customers. These investments differ widely in their financial and economic risks. Large, long-lead-time supply projects have more uncertain returns than relatively safe investment in transmission and distribution. Among electric generation projects, there can be substantial differences in risk (Ford and Yabroff 1978; Kahn 1979). The conventional analysis fails to capture these differences. This failure is the subject of concern within the utility planning community; it was discussed recently in a committee paper sponsored by the Power Engineering Society of the Institute of Electrical and Electronic Engineers (IEEE) (Platts and Womeldorff 1979).

Roughly speaking, projects with greater risk ought to return a greater proportion of their investment annually. One framework in which to assess this tradeoff between risk and return is the capital asset pricing model (CAPM) (Sharpe 1970; Mossing 1973). In the case of electric-utility investment, the risks associated with end-use substitution investments such as solar hot-water heating or ceiling insulation appear considerably lower than those associated with large-scale generation projects (Kahn and Schutz 1978). If correct, this means that the return required from such investments ought to be lower than the weighted-average cost of capital. This will mean a lower effective interest rate for solar financing by the utility. Relative risk is explored further later.

If we assume that the risk of solar investment by utilities is sufficiently low to justify an interest rate which is lower than the pretax cost of capital, it remains to discuss the regulatory devices available to capture this effect. As a practical matter, it would be possible to roll in solar investments under capitalization and charge them at a cost of capital which is less than the pretax rate. This amounts to changing the capital structure on the margin, weighting it more heavily toward low-cost instruments and less toward common equity. Under a financing arrangement the treatment would be essentially the same, although it would be more transparent that this class of investment is being handled differently from conventional investment. A financing subsidiary, for example, might be capitalized at 10 percent or 20 percent common equity, and the rest would be debt. The effect of capital structure on pretax cost of capital is shown in table 5–20.

A substantial administrative problem associated with financing plans is the design of the repayment schedule. This is especially important when the effect of social mobility on the term of loans is considered. Undercapitalization return on the investment is achieved as part of the ordinary rate-making process. Since the utility "owns" the equipment, it does not matter if the nominal occupant of a building with this equipment changes. The current occupant of the residence will still make "payment" through the rate structure. Where an explicit loan is made, some provision must be made for solar borrowers who move from their solar residence. Is the loan liquidated at this time or transferred to the new owner? What if the new owner does not want to assume the loan? This problem is significant because the average turnover time for houses is less than the 20-year amortization often required to make solar loans cost-effective. It is estimated that the average house changes owners at a point between the fifth and tenth year from purchase (Testimony of T.A. Keefe 1979). Few solar projects are cost-effective if amortized at 10 years or less. Thus, not only is there uncertainty over the economic life of solar systems, but also demographic mobility tends to reduce and complicate one of the main advantages of public-utility financing—the ability to raise long-term capital. The most

Table 5–20
Capital Structure and Average Cost of Capital

Ratio		Incremental Cost (%)	After-Tax Weighted Cost (%)	Tax Multiplier [a]	Pre-Tax Cost of Capital (%)
Standard Case					
Debt	50%	9.5	4.75	1.00	4.75
Preferred stock	10%	9.5	0.95	2.04	1.94
Common equity	40%	14.0	5.6	2.04	11.42
			11.30		18.11
Leveraged Subsidiary					
Debt	80%	9.5	7.60	1.00	7.60
Equity	40%	14.0	2.80	2.04	5.71
			10.40		13.31

[a]Calculation of tax multiplier:

1. Reduce income by state income tax rate (9%): $100 - 9 = 91\%$.
2. Calculate federal tax at 46%: $91\% \times 46\% = 41.86\%$.
3. Add state income tax: $41.86\% + 9\% = 50.86\%$.
4. The tax multiplier is the reciprocal of 1 minus the marginal tax rate, or $1/(1 - 0.5086) = 2.04$.

outstanding practical solution to this dilemma is the energy-conservation financing plan designed and implemented currently by the Pacific Power and Light Company. It is to this subject that we now turn.

The Pacific Power and Light (PPL) Company Residential Energy Efficiency Rider. The PPL is an investor-owned electric utility operating principally in Oregon, but in six other states as well. Its residential energy efficiency rider is a unique combination of capitalization and financing elements used to encourage investment in residential weatherization. Although not addressed to solar applications, the approach is generalizable under certain conditions.

The main features of the PPL program are as follows:

1. PPL performs a home energy audit and recommends specific weatherization investments whose life-cycle cost is less than the marginal cost of new supply.
2. Upon approval of the homeowner, PPL arranges for contractor installation of the weatherization materials.
3. PPL pays for all materials and labor.
4. The homeowner agrees to repay these original costs with no interest on or before the point of sale.
5. PPL accounts for these investments by adding them to the rate base, using no amortization.
6. All customers pay the carrying charges on the capital for as long as the loan is in the utility rate base.
7. Upon transfer of the home and repayment of loan, the rate base is reduced by the amount of the loan.

This program is attractive to all parties involved in the transaction. Customer response has been good; a substantial backlog of requests for participation has already developed. The current completion rate is about 5,000 homes per year (Letter from C.P. Davenport 1979). The scale of the program is sufficiently large to support the assertion that public-utility financing can make major differences in the adoption rate of weatherization investments. Benefits of this program to the utility are discussed in some detail below.

The main structural innovation of the PPL plan is the use of the time of property transfer as the point at which the loan must be liquidated. This feature, combined with the capitalization of the loans in rate base, has the effect of evening out the allocation of program costs among participants and nonparticipants. Under simple capitalization, in plans such as the FEA's Rosenberg proposal (ICF, Inc. 1977), residential conservation investments were to be capitalized in rate base for their estimated economic lives.

The FEA proposal used 15 years for this lifetime. This meant that nonparticipants carried the cost of the program over the entire period. Under the PPL plan, the nonparticipants' burden will end long before the benefits of the investment cease. Since the PPL plan is still an actual loan, where all customers bear the interest cost, repayment on resale eliminates a basic inequity of simple capitalization. Nonparticipants do not continually pay for the benefits received by others. While there are still questions of customer equity involved in the PPL plan, its combination of features tends to eliminate some of the most troublesome features of simple financing or capitalization.

Apart from its structural innovations, the PPL plan has a particular definition of cost-effectiveness used to evaluate end-use conservation investments that is a major constraint on program scope. Conventional utility economic analysis of investments for central station supply is based on the minimum-cost criterion. That alternative is best which has the lowest marginal cost. To account for the differing incidence of costs and benefits to participants and nonparticipants, PPL has proposed a more stringent criterion on its program. An end-use conservation investment program must save energy at an average cost which is less than the difference between the utility's marginal cost of new supply and the current average retail cost. If a program meets this test, the nonparticipants will have no higher a cost of energy under the program than without it. The derivation of this criterion is given below, and its application is discussed in the California case study presented later.

Leasing Arrangements. Leasing capital equipment, rather than purchasing it, is a financial device introduced to transfer tax benefits among parties to a transaction, so that all actors are better off (California PUC Energy Conservation Team 1977). It has recently become a factor in public-utility financing. San Diego Gas and Electric Co., for example, sold its Encina 5 power plant to the Bank of America and leases it back from them. The arrangement resulted in a net cost of capital to the utility of about 6 percent. While this is an attractive rate of interest in today's market, the long-term effect on the utility's credit is not positive. The reason is that utility bond rating agencies view the lease as a long-term debt obligation which leverages the utility further and provides no equity protection ("San Diego's Utility Typifies Industry Woes," *Business Week,* May 28, 1979).

The San Diego Gas and Electric Co. lease is based on a situation in which the utility has federal tax credit that it cannot absorb because of insufficient revenue. These benefits are passed through to the bank which shares the benefit in the form of a lower interest rate. Other tax situations are possible. The natural-gas utilities are not generally in the same tax position as electric utilities or combination companies. Electric power generation is so capital-intensive that electric-utility investments generate substantial tax preferences. Natural-gas utilities, on the other hand, have relatively smaller capitalization. Their construction projects are either smaller than those of electric utilities or so large in nature (liquified natural gas, for example) as to require wholly unconventional financing. For relatively modest-scale incremental investments, gas-utility solar financing using leasing techniques would enable the utility to capture tax benefits not otherwise available to it and pass some of these along to customers. In a study of gas-utility financing alternatives for residential solar applications, MITRE found the leasing alternative most attractive (Boyd et al. 1978). This conclusion followed from assumptions of more highly leveraged utility subsidiary financing than under simple capitalization and the capture of tax benefits. It is as yet an unresolved issue whether utilities which own or lease solar equipment would actually qualify for conventional investment tax credit. Under utility leasing there would be no capture of state or federal tax credits aimed at consumers.

Utility Solar Subsidiaries. Public-utility companies sometimes engage in businesses that are not part of their monopoly franchise, but which may be tangentially related to their main activities. To separate these nonutility operations from the regulated activities, it is conventional to create subsidiary corporations for nonutility businesses. For particular activities it may not be entirely clear whether it does or does not come under the scope of the monopoly franchise. In these cases, subsidiaries are also useful devices to create a financial separation from the parent company. Such a separation may be used to allow more latitude to the subsidiary than the parent or, conversely, to allow a close regulatory scrutiny of the particular activity.

One of the major concerns involved in organizing a utility subsidiary is the determination of an appropriate capital structure and accounting correctly for the cost of a subsidiary's capital (Marx 1978; Jones and O'Donnell 1978; Seeds 1978). Table 5-20 indicated that capital structure has a major impact on the average cost of money. What is less clear is the justification of different capital structures and the imputation of costs to each instrument. The cost imputation is complicated, in turn, by the variety of corporate devices which can be used to control the subsidiary.

The most logical grounds on which to impute subsidiary capital structure and costs are on the basis of project risk (Sussman 1979). The practical problem is that usually the risks of a new project are not readily quantifiable beforehand. Some general guidelines with regard to the effects of diversification are available. In a substantial empirical study across many industries, Rumelt found that a limited amount of diversification could reduce the risk of parent corporations (Rumelt 1974). However, unless it were constrained to some functional relation to the main line of business, diversification would show no particular benefit. In the public-utility sector, Fitzpatrick and Groebner (1978) found confirmation for these general conclusions. In particular, natural-gas utilities which have diversified widely into unrelated businesses appear to have *increased* their risk by such activity. This increases the cost of capital to the parent utility's customers. On the other hand, electric utilities have relatively little non-utility activity and could, by some limited diversification, reduce their risk. The specific risks of utility solar investment are discussed below.

The results of Fitzpatrick and Groebner suggest the third major issue associated with utility subsidiaries—whether these businesses should be regulated. This decision often will be made on the pragmatic ground of whether a would-be regulator has sufficient staff time and resources available to regulate subsidiaries. If such time and resources are not available and the risks to utility customers appear substantial, then the regulators' only option is to forbid the activity. More ambiguous situations arise when the risks are not well understood. For the issue of solar investment by utilities, a leader-follower situation among state regulators is likely to develop. In California, substantial regulatory analysis of the issue is currently being pursued. This process is likely to generate information and perhaps precedents for other commissions to rely on. States with limited resources for regulatory scrutiny may be expected to develop guidelines based on California's experience.

Special Bonding Authorities. The last major feature of a utility solar financing program to be examined here is the use of special bonding authorities as a means of raising relatively low-cost capital. With municipalities and other specially constituted local agencies raise capital, they sell bonds whose interest is tax-free to the purchaser. The interest paid on the bonds of private corporations, including investor-owned utilities, is taxable. Therefore, the latter will have a cost of debt capital which is greater than the tax-exempt debt sector. This fact has created interest in the possibility of financing residential solar systems through tax-exempt mechanisms.

One approach to the tax-exempt capital market is through existing publicly owned utilities. A widely cited example is the cit of Santa Clara, California. The city currently leases solar swimming pool heaters to residents through its water department. There are plans to lease solar water heaters (Southwest Energy Management, Inc. 1978). The use of municipal utilites as a vehicle for widespread implementation of solar systems may be attractive where these utilities have established service territories and are in sound financial condition. If such institutions must be established as a precondition for utility solar financing, then major advantages of utility finance—its security and convenience—will be missing.

Special bonding authorities also may be used for access to the tax-exempt capital market. In California, the state government administers a Pollution Control Financing Authority which issues tax-exempt bonds to finance investments in pollution control. In the past, investor-owned utilities have used such funds to finance power plant scrubbers (California Pollution Control Financing Authority 1979). It has been suggested that such an arrangement might be used for utility solar financing (Testimony of T.A. Keefe 1979). It is not clear that such an arrangement would qualify under the legislation. Furthermore, in this particular case, there are limits on the amount of capital obtainable through this mechanism. The Authority is legally capable of floating $50 million per quarter. In the five years since its inception, total funding has been about $270 million (California Pollution Control Financing Authority 1979). If the roughly 400,000 electric water heaters in California were replaced at a cost of $2,500 each, the total capital requirement would be about $1 billion. This is almost four times the amount of bonds issued. At the maximum rate, it could be financed over five years, but this would crowd out any other investment in pollution control.

Economic Impacts of Solar Investment on Utilities

In this section a survey is made of the various economic effects of utility investment in on-site solar on the utility company involved. The discussion addresses both the planning process for new conventional utility supply and the current financial position of the utility industry. Special consideration is given to the role of the federal tax preferences. Relatively little attention is paid to the specific program features identified above in hopes of concentrating on the fundamental choices involved in determining whether the utilities ought to play a role in solar energy financing.

Internalization versus Externalization. If regulated utilities are not allowed a role in on-site solar finance, they will still be impacted intermediaries as the residential solar market develops. In a scenario where utilities are external to the solar market, the main policy questions of interest center on the ability of the utility to respond to that market development. The appropriate responses would involve reoptimization of the utility supply plans to reflect the changed nature of demand facing the utility.

Literature analyzing the solar/utility interface usually is based on an implicit view of this adjustment process. Bright and Davitian (1978), for example, assume in their study of solar backup energy costs that all changes in utility demand caused by solar penetration in the residential market are known with certainty. Therefore, costs can be calculated by comparing various runs of a utility optimization model. At the other extreme, Willey (1978) analyzes several scenarios involving large-scale solar market development where the utility either capitalizes or ignores on-site solar. The latter study finds that utility capitalization of solar results in lower utility costs than the case where the solar market develops and the utility makes no adjustment whatsoever.

It is likely that reality lies somewhere between the assumptions of perfect information and no adjustment process at all. Another way of putting this is that, from the utility perspective, uncertainty is inherent in the planning process. If the utility is external to the solar market development process, then that process will compound the already substantial demand uncertainties facing both electric and gas utilities. While utility planners are beginning to recognize the need to treat forecasted demand growth probabilistically (Platts and Womeldorff, 1979), the current state of the art shows major unexplained structural differences among the existing demand forecasting models (Cherry 1979).

Utility solar financing would help make the solar market development process internal, rather than external, to utility planning. In this case, discriminatory solar rates would be less likely to be proposed by utilities and adopted by regulators. In theory, an integrated utility planning process would choose among solar, conservation, and conventional technologies on an unbiased economic basis. Thus, the utility would no longer have an incentive to defend its economic stake in large supply projects whose demand would be less expensively served by solar investment. Since internalization carries with it the risk of monopoly action in the solar market, and the concurrent danger that technology would be retarded, some steps short of utility ownership deserve consideration. These alternatives are discussed below.

Electric Utility Financial Risk Profile. Any proposal for utility solar financing must examine the impact of such schemes on the risk structure of the utility. For this assessment to be realistic, it is important to understand the current financial position of the utility industry. By general consensus, the outlook for electric utilities is not particularly good ("A Dark Future for Utilities," *Business Week,* 1979). The major factors contributing to the industry's problems have been alluded to above. The cost, scale, and construction time required for major new supply projects have been growing. This has been coupled with uncertain demand growth that has lagged behind past expectations. The interaction of cost escalation, long project lead times, and softening demand have combined to put a serious strain on electric-utility cash flow (Kahn 1979b).

From the perspective of the utility's financial stability and viability, investment in on-site solar involves a tradeoff between technical risk and the flexibility of small-scale incremental supply. In a fundamental way, on-site solar resembles nuclear and hydro-power generation in that all these technologies are substitutions of capital for conventional fuels. In a regulated industry, such substitutions are advantageous because they immunize the utility's earnings from the effects of regulatory lag. The current climate of rising marginal costs and persistent inflation tends to cause earnings attrition, because utility rates are typically set on the basis of cost estimates that turn out to be less than actual costs. Fixed costs by nature are not subject to inflation or escalation once the initial capital has been sunk. In a regulated industry, the adjustment of variable costs to inflation and escalation will always lag as a result of the administrative delays attendant on the rate-making process. The principal advantage of solar investment as a substitute of capital for fuel is the small scale of each unit.

Nuclear generation exhibits diseconomies of scale that are reflected in the standard financial ratios used to evaluate a utility's corporate credit. The ratio of earnings to interest payments, measured in various ways, indicates the extent to which a bond holder has assurance that he will be paid. Bertschi has shown a systematic relationship among these ratios which distinguishes companies building nuclear plants from those which have no nuclear construction (Testimony of R.L. Bertschi 1979). The capital requirements for a nuclear plant are of such a magnitude and occur over such a long time that a severe strain is placed on the credit of their sponsors. Once construction is complete, this strain disappears and the financial stability of the utility improves.

Solar investments, while capital-intensive, would be made in increments that are more easily adjusted to the financial capability of the utility. This benefit is magnified by the short lead time involved in most solar residential applications. It is the long construction and licensing period for large-scale projects which imposes the financial strain. Under the most common regulatory procedures, the utility will not earn a return on capital allocated to construction until the plant goes into service. Although there are regulatory remedies to the financial lag induced by long construction periods, these are not politically popular in many constituencies (Holt, 1979).

It should be emphasized that the financial strains and risks of large-scale projects are reflected in the capital market. One measure of the capital-market risk evaluation is the differential bond yield on public-utility debt issues. This shows that market risk premiums are higher than they have been historically, except during the Great Depression. For the last few years, studies have also shown a risk premium in the common-equity market that is linked to the magnitude of construction activity (Benore 1978; Fitzpatrick and Stitzel 1978).

Public-Utility Investment Choice. One crucial consideration in public-utility financing proposals is their impact on utility investments. In this section the case is made that utility investment in solar energy technology, either by financing consumers by ownership, would be advantageous to the utility. This conclusion is drawn primarily from the works of Kahn and Schutz (1978) and Willey (1978).

The economic incentive for utility involvement in the solar energy industry concerns the potential risk and return for solar investments relative to conventional investments (such as coal and nuclear for electric utilities). As previously mentioned, Kahn and Schutz have pursued this question using the capital-asset-pricing-model (CAPM) to estimate the market-oriented risk associated with on-site investments (by capitalization or financing) and conventional generation investments. The CAPM, as defined by Kahn and Schutz, is based on the economic theory that the capital market is characterized by a relation between the risk and return on the investment. This basic theory is utilized by CAPM as follows: "CAPM assumes that all investments can be thought of as random variables characterized by an expected value of the return and some variance. The variance of the return is identified as the risk of the investment" (Kahn and Schutz 1978, p. 4). This can be expressed algebraically, using Kahn's notation and model, as

$$E(R_i) = R_f + \beta_i[E(R_m) - R_f] \quad \text{where} \quad \beta_i = \frac{\text{cov}(R_i, R_m)}{\text{var}(R_m)}$$

where R_i = return on a risky security i

R_f = risk-free rate, as on the return for short-term Treasury Bills

R_m = average market return for a portfolio

β_i = "volatility" of security i's return

Now, β_i can be called the systematic risk measure which depends on the degree to which the asset return covaries with the average return on all assets, that is, the economy as a whole. In the right side of the above equation, $\beta_i[E(R_m) - R_f]$ can be expressed intuitively as the "Risk Premium which is proportional to the 'volatility' of the security and the market price of risk" (Kahn and Schutz 1978, p. 6), for $E(R_m) - R_f$ is simply the return that will be earned on the average market portfolio to compensate the investor for the risk of his investment over the risk-free rate R_f. The parameter β_i reflects the sensitivity of R_i to market fluctuations. If $\beta_i = 1$ for some asset i, then $E(R_m)$ will equal $E(R_i)$, which implies that the return on i will be the same as the market return. If β_i is greater than 1, then the asset will vary more than the market average m. Kahn and Schutz note that growth stocks with "out-perform the indices" are indicative of a high beta (β). Utilities, however, characteristically exhibit betas that are less than 1, indicating that they are less sensitive to market fluctuations than the average portfolio.

Utilizing this analysis of the risk of various investments can be of importance in determining the desirability of on-site solar investments relative to conventional investments for the utility. Thus it would be helpful, given the CAPM, to determine the betas for on-site solar investments relative to conventional generation. If the beta for solar is lower, one might expect utilities to want to invest in the industry through capitalization or financing.

Kahn and Schutz provide estimates for betas of utilities, indicating that they are indeed low beta stocks. Beta for utilities is calculated simply from a regression model of past values of R_i on past values of R_m. Kahn and Schutz refer to this as the market model with the following regression equation:

$$R_i = a_i + \beta_i R_m + e_i$$

For utilities overall, according to Standard and Poor's utility index $\beta = 0.91$; for San Diego Gas and Electric Co., $\beta = 0.67$; for Pacific Gas and Electric Co., $\beta = 0.56$. These figures indicate that utilities have historically been very stable, reflecting the stability of investment in conventional generation technologies. It should be noted, however, that these calculations of beta for the electric utilities are based on past returns and investments. If present and future expectations are considered, then the betas for conventional generation are substantially higher.

Kahn and Schutz provide data for Pacific Gas and Electric Co. to determine the effect of marginal components (new plants) to their production process on estimates of beta. In considering the beta for total capital, including planned conventional plants (which will be 56 percent of existing equity by 1996), Kahn and Schutz estimate $\beta = 0.91$. When this is broken down into historical and marginal betas, they estimate that the beta for marginal plant is 1.36, considerably higher than the $\beta = 0.56$ for historical plant investment.

The estimate of beta for on-site solar installations is also a relatively simple process. Because of several factors that are discussed below, the risks involved in on-site solar investments appear to be very minimal. The major risk is the duration of the capital investment. Solar systems represents investments that are fairly permanent and unchanging for several decades; thus there is always the risk that better uses for the capital invested in solar will be found. As one looks at capital markets, one finds that this risk is often expressed in higher rates for long-term investments. Thus one method for determining beta in the case of on-site solar is to consider the beta assets that have comparable risks. Kahn and Schutz utilize the Standard and

Poor's high-grade corporate bond index to estimate beta for solar installations. They conclude that the beta for solar installations is approximately 0.15.

Kahn and Schutz indicate that the only risk that is unique to solar investments is the default risk that comes from people refusing to pay their gas or electricity bills or vacating the house without payment. This can result in a default risk for solar investments in residential installations nor currently experienced in conventional investments.

There are, however, several other risk measures that solar does not share with conventional investments which result in a low beta for on-site solar and a high beta for conventional generation. First, there is significant demand uncertainty for conventional plants; solar installations are not sensitive to uncertainties about long-term demand because they are matched to end-use demand. Second, there are changes in energy cost which add to the risk of conventional systems; solar energy investments are immune from this uncertainty because they require no noncapital inputs that are likely to affect the cost of solar energy, nor are there any environmental regulations likely to affect cost. Third, there is the risk that the utility has invested in a "lemon" technology, that is, a technology that proves to be inadequate because of some inherent design or conceptual problems. This is certainly the case with nuclear plants, whose future is filled with uncertainty. While solar is also a new technology (and even a potential lemon), the lead time for solar installations is short, and they can be invested in in small quantities so that the overall risk is small.

Willey has done a study of alternative energy systems for Pacific Gas and Electric Company in California and concluded that utility investment in solar energy was actually a more sound investment than conventional technologies. He compares three potential energy systems, each of which will provide the projected needs for the PG&E district. The conventional system, Ei, is characterized as "the current resource plan for energy investments during 1978–1996, which includes two nuclear units (1200 MW each) and two coal units (800 MW each) assumed to become operational during the 1980's; and six 'base load' (800 MW each) coal units or four nuclear units (1200 MW each), assumed operational during the 1990's (Willey 1978, p. A69). The second alternative energy system, and the one most significantly different from the conventional system, E3, is characterized as "an investment plan which eliminates nine out of ten coal and nuclear units (leaving one 800 MW unit, now planned by PG&E for operation by 1996) and incorporates somewhat extensive development of alternative sources, including wind reaching the 'moderate development' level by 1996" (Willey 1978, p. A70). Alternative sources include on-site solar space- and water-heating installations, cogeneration facilities, geothermal, wind, and increased end-use efficiency.

Willey concludes that alternative energy sources are cost-comparative with conventional energy sources and that investment in alternative technologies is a sound choice. The basic element in his investment choice analysis is that there is greater financial cost associated with technologies that require large amounts of capital and have long lead times. A crucial assumption is that the utility will be able to incorporate solar investments into the rate base and reap any tax benefits. Given that alternative energy sources have shorter lead times, reduced construction costs, and lower fuel costs, Willey concludes that "compared to the PG&E system [E1], E2 and E3 experience (i) higher cash flows, (ii) higher internal to total finance ratios, (iii) significantly reduced common share issuances, (iv) dramatically lower AFDC, (v) higher cash interest coverage ratios, and (vi) approximately the same book value per common share..." (Willey 1978, p. A90). If we translate these advantages into financial savings, "the 1978–1996 cumulative future earnings available for common resulting from system E1 and system E3 discount (using a rate of return on equity of 0.150) to 1978 present values of $2782 million and $3254 million respectively" (Willey 1978, p. 13). Considering the extra earnings available from E3 investments, PG&E would be losing $75 million per year of earnings in 1978 present value by investing in the E1 system.

Therefore it appears that, based on the Willey and Kahn and Schutz analyses of utility investment in alternative technologies, public utilities do have incentives to invest in on-site solar installations. In terms of public-utility financing schemes, it would be advantageous for the utility to support such proposals, given the costs associated with long-lead-time technologies in the face of demand uncertainties. There is a major exception: construction commitments to conventional plants foreclose this option.

Without substantially more experience with widespread use of residential solar technology, it is not possible to dismiss the technical risk and uncertainty associated with any relatively unconventional technology. Therefore, as a practical matter, any utility solar financing program ought to start at a relatively small scale and grow larger as more performance experience is acquired. Although, in principle, utilities ought to be able to provide maintenance services for solar investments, it might be more desirable for these costs to be borne by participants in utility financing programs. Such a treatment of maintenance expenses would tend to minimize the technical risk of the program to the utility. Again, more actual experience will indicate the dimensions of this potential problem.

Federal Tax Effects—Excess Investment Tax Credit. The role of federal corporate income taxes in determining a utility's cost of capital is indicated in table 5–20. The nominal income tax rate of 46 percent, however, is usually offset by tax preferences associated with capital investment. The two major tax incentives for utility investment are accelerated depreciation and investment tax credit (ITC). Given the size of current electric-utility capital programs, the effective tax rate for utilities ranges from 0 to 20 percent (Morgan 1976). This effective rate would be even lower on average were it not for a limitation on use of ITC. In the tax revision laws of both 1975 and 1978, explicit limitations were placed on the use of ITC to offset tax obligation. These limits vary from year to year, going from 70 percent in 1979 to 80 percent in 1981 and 90 percent in 1982 (Mulligan, undated CPUC memo). The importance of this limitation is that many utilities are currently in the anomalous position of having substantial ITC carry-forward balances that cannot be used (Smartt 1979). The constraint which creates this is the inability of the utility to generate sufficient income to absorb the credits. The importance of this effect is that it can create a de facto tax credit for utility solar financing. Such a program, or indeed any program which generates revenue, will capture some of the excess ITC. This will lower the incremental tax rate on such programs in a significant way.

It is instructive to examine some data on excess ITC. In one recent survey of forty-five investor-owned utilities, 22 percent were found to have ITC carry-forward balances which averaged $18 million (Johnson 1979). It is not surprising that Pacific Power and Light (PPL) has an ITC carry-forward of considerable proportion. According to its 1978 annual report to the California PUC, PP&L had about $12 million in excess ITC. It is unlikely that this balance will decline. This is due to the magnitude of the PP&L construction program. Over the next seven years (1979 to 1985), PP&L's capital budget for generation and transmission projects alone is estimated at $1.7 billion (Testimony of D.W. Sloan 1979). This will generate approximately $170 million in ITC. The average ITC over this period would be $24,4 million per year. In 1978, PP&L used a little over $21 million in ITC to offset income taxes (1978 Annual Report of PP&L). If we consider that additional ITC is likely to be generated by investment in distribution plant, PP&L can reasonably look forward to a positive ITC carry-forward balance into the mid-1980s.

The effect of utilizing excess ITC on the incremental cost of capital can be seen by recalculating the tax multipliers used in table 5-20. For illustrative purposes, let us assume that the average ITC utilization limitation is 80 percent. Table 5-21 retraces table 5-20 calculations of marginal tax rate and pretax cost of capital. The calculations in table 5-21 show that the impact of excess ITC on the cost of capital is large. In the case of a standard capital structure, the effective cost goes from 18.11 to 12.68 percent. For a leveraged susidiary, the cost goes from 13.31 to 10.99 percent.

Thus, while the phenomenon of excess ITC has significant implications for utility solar financing, it is not particularly clear why some utilities have

Table 5–21
Effect of Unutilized Investment Tax Credit Marginal Cost of Capital

Tax Multiplier

1. Reduce income by state income tax rate (9%): $100 - 9 = 91\%$.
2. Calculate federal tax at 46%: $91\% \times 46\% = 41.86\%$.
3. Net out ITC up to 80%: $(1 - 0.80) \times 41.86\% = 8.37\%$.
4. Add back state income tax: $8.37\% + 9\% = 17.37\%$.
5. Tax multiplier $= 1/(1 - 0.1737) = 1.21$.

	Rate(%)	*Incremental Cost (%)*	*Weighted Cost (%)*	*Tax Multiplier*	*Pretax Cost of Capital (%)*
Pretax Cost of Capital: Standard Case					
Debt	50	9.5	4.75	1.0	4.75
Preferred stock	10	9.5	0.95	1.21	1.15
Common equity	40	14.0	5.6	1.21	6.78
			11.30		12.68
Pretax Cost of Capital: Leveraged Subsidiary					
Debt	80	9.5	7.60	1.0	7.60
Equity	20	14.0	2.80	1.21	3.39
			10.40		10.99

significant ITC carry-forward balances and others do not. The most likely explanation is the different state regulatory treatment of construction expenditures and tax preferences. A more systematic investigation of the relation between regulatory practices and ITC carry-forward is conducted later.

Economic Impacts of Utility Solar Financing on Customers

Utility solar financing raises a variety of issues regarding the equal treatment of participants in such programs as opposed to that of nonparticipants. To make programs attractive to customers, utilities will make inducements whose costs may or may not be justified. Relatively straightforward tests may be applied to assess the equity among utility customer classes of solar incentives. The issue becomes more complicated when the incentives of a utility solar financing program interact with other incentives such as tax credits. Here the remedy for inequity is less transparent. Finally, there is a range of economic equity questions arising from the recognition that utility solar financing may not be society's least-cost alternative. The social cost perspective is explored in detail below.

Nonparticipant Break-Even Requirement. Utility investment in end-use efficiency differs fundamentally from investment in centralized supply because the benefits of the former have more unequal incidence than those of the latter. In principle, no single class of customers would benefit more from a new power plant than any other class. In practice, there may be rate-making devices which distort the benefits of new investment to favor one class (Testimony of Eugene Coyle before the New Jersey Board of Public Utilities 1978), but there is nothing inherently unequal about the distribution of benefits. Where end-use efficiency is concerned, however, the benefits to participants are immediate and substantial in the form of reduced consumption and lower utility bills. The nonparticipant receives the indirect benefit of decreased requirement for new high-cost supply projects. Not only is this less tangible than a reduced utility bill, but also it is possible that nonparticipants could bear an increasing share of utility revenue requirements. This would mean that their average cost of energy was higher because the total revenue collected form participants had diminished.

To avoid this potential inequity, a bound can be derived on the incentive to participants which will avoid increasing the average cost of energy to nonparticipants. Essentially, the appropriate incentive should be the difference between marginal and average unit energy costs times the amount of energy displaced by conservation or solar investment. This incentive can be implemented through rate structures in the case of no utility solar financing (IFC, Inc. 1978). Alternatively, the criterion can be used to set cost goals for a utility capitalization program for end-use substitution investments. This is exactly the approach of Pacific Power and Light.

A formal derivation of the break-even cost for nonparticipants is given below. The presentation follows a simple model used by PP&L (Testimony of J. Shue 1978).

Let G = initial consumption of nonparticipants
$\quad C$ = initial *total* consumption
$\quad I$ = marginal cost of supply per kWh
$\quad g$ = annual growth rate
$\quad x$ = cost of conservation (or solar) per kWh
$\quad r$ = average cost of supply per kWh initially

Assumption: All growth in load is from *non*participants in a conservation program (under conservation *total* consumption is constant before and after conservation measures).

1. gC = new load, supplied by plant at marginal cost

Nonparticipant revenue requirements = proportional share of total dollar requirements

Proportion = $\dfrac{\text{initial nonparticipants' consumption + new load of nonparticipants}}{\text{new } \textit{total} \text{ load}}$

$= \dfrac{G + gC}{C + gC}$

Total dollar requirements = $rC + I(gC)$

Nonparticipants' share of revenue = $\left(\dfrac{G + gC}{C + gC} \right)\left(rC + IgC \right)$ (5.1)

2. gC = new load, "supplied" by conservation

In this case, the proportion of total supply used by nonparticipants *increases:*

Proportion = $\dfrac{G + gC}{C}$ (only C because there is no new supply for the system overall)

Total dollar requirements = $\underset{\text{(initial)}}{\downarrow rC} + \underset{\underset{\text{(rate)}}{\downarrow \text{(amount)}}}{x(gC)\downarrow}$

Nonparticipants' share of revenue = $\left(\dfrac{G + gC}{C} \right)\left(rC + xgC \right)$ (5.2)

3. If revenue from *non*participants is to be the same under the conservation approach as under new-supply (at marginal cost) approach, then equation 5.1 can be set to equation 5.2:

$$\left(\frac{G + gC}{C + gC} \right)\left(rC + IgC \right) = \left(\frac{G + gC}{C} \right)\left(rC + xgC \right)$$

then $xg\cancel{C} = \dfrac{\left[(\cancel{G} + g\cancel{C})/(\cancel{C} + g\cancel{C}) \right](r\cancel{C} + Ig\cancel{C}) - r\cancel{C}}{(\cancel{G+gC})/C}$

$xg = \left(\dfrac{1}{1 + g} \right)\left(r + Ig \right) - r$

$$xg = \frac{r + Ig - r(1 + g)}{1 + g}$$

$$= \frac{(I - r)g}{1 + g}$$

$$x = \frac{I - r}{1 + g}$$

If $g < 1$,

$I - r$ = difference between marginal cost and average cost (initial)

Interaction of Utility Solar Finance with Other Solar Incentives

A number of incentives for development of the residential solar market currently exist or are proposed. Where these simply compete with utility solar finance, there is no particular policy problem. Society may wish to favor one kind of financing over another, but there is nothing extraneous which complicates the choice. Other incentives will interact economically with utility solar financing, and this creates policy complications. The main difficulty occurs with federal income tax credits for individuals. Before this case is explored, it is convenient to take up a less difficult case, the interaction of utility rate reform with utility solar financing.

The Public Utility Regulatory Policies Act (PURPA) mandated the explicit analysis of electric utility rate reform by state regulatory commissions. Such reforms might have an explicit or implicit incentive effect on the residential solar market. For example, time-of-day rates based on existing or projected daily cost variations could favor residential solar applications for hot-water heating in a summer-peaking utility. A cost study of this problem which considered the solar alternative explicitly would, in all likelihood, come up with a solar incentive that would be more attractive than the implicit incentive which would result from no solar analysis at all. For example, a recent study of this problem concluded that discounts to solar users were appropriate if limited to the difference between marginal and average costs (ICF, Inc. 1978). Since this is the same criterion underlying the PP&L zero-interest loan program, it would be unfair to allow both the discount and the favorable financing. This would be the same as giving the justifiable subsidy twice. In principle, such difficulties are avoidable, since the utility is internalizing all costs and can be expected to avoid excessive incentives. In practice, the possibility of utility solar financing may well complicate the process of rate reform under PURPA. The appropriate assumptions for cost studies of rate reform depend on the policy toward utility solar financing. If this policy is changeable or unknown, then the accuracy of rate-reform cost analysis becomes questionable. Resolution of such problems amounts to the formulation of consistent state regulatory policy. In principle, this is feasible.

The interaction of utility solar financing with the tax-credit incentives is more complex. The equity problem is simple to describe: excessive incentives. The resolution is more difficult because there is no institutional framework for rationalizing and coordinating justifiable subsidies from the perspective of utility costs with those justified by social costs. In practical cases, it may turn out that the tax-credit mechanism is literally being used twice under utility solar financing. The excess ITC situation described above can turn out to be a significant determinant of costs. Thus, a participant in such a solar financing program would be eligible for both zero-interest loan and substantial state and federal tax credits. To avoid this double incentive, it has been proposed that the state credit be signed over to the utility (Testimony of T.A. Keefe 1979). This solution would have administrative complexities and does not really solve the problem when excess utility ITC is involved. An alternative would be to terminate tax credits for participants in utility solar financing programs.

The role of the utility in the solar energy industry from any angle is a complex issue. There are a number of institutional economic problems associated with the issue aside from those already mentioned. Chapter 6 examines some relevant questions and proposes several novel solutions.

Notes

1. The load criteria are for the most part incorporated in the federal Public Utility Regulatory Policy Act (1978) guidelines.

2. Personal communication.

3. Kelly et al. (1977) assumed that the utility plant mix was optimized to meet new load patterns caused by different solar technologies. The optimization procedure is not specific. Bright and Davitian (1979) have also found this to be the case for space heating for LILCO and Public Service of New Mexico.

4. See the testimony of S.P. Reynolds in OII42, California Public Utilities Commission (CPUC), Los Angeles, October 30, 1979.

5. This section was written by Charles Cicchetti and William Gillen. For an expanded discussion of marginal-cost pricing, see Berlin, Cicchetti, and Gillen (1974) and Cicchetti, Gillen, and Smolensky (1976).

6. The actual tax credit passed by Congress in October 1978 lowered the benefits for energy conservation to 15 percent of the first $2,000. The percent effects of this alteration are noted accordingly.

7. The difficulties and advantages of utility involvement with decentralized solar energy systems are extensively reviewed in Feldman and Anderson (1976b), Federal Trade Commission (1978), and Laitos and Feuerstein (1979).

8. Ray Czahar *A Study of Solar Financing,* California Public Utilities Commission, April 24, 1979, chap. 2.

Public Utilities and Solar Energy Development— Institutional Economic Considerations

Roger G. Noll

Since the Federal Power Commission (FPC) began regulating the field price of natural gas that is sold in interstate markets, supplies of gas have dwindled rapidly (Pindyck 1974). The opportunities available to gas utilities for increasing their fuel supplies—primarily imported liquified natural gas and gasified coal—involve sufficiently high costs that, for some uses, solar technology for space and water heating now stands on the brink of economic viability when used in tandem with gas. More recently, regulatory interventions using administrative mechanisms rather than prices to allocate the available gas threaten to cut off some gas users regardless of their willingness to pay. For these users, solar heating holds the promise of being the next best alternative to gas, especially if the alternative energy source must avoid serious environmental degradation.

Costs have also risen more rapidly than the overall rate of inflation in the electric industry. Rising fuel prices, while troubling, are only a relatively small part of the story. Increasingly stringent environmental standards have added several hundred dollars per kilowatt to the costs of coal, oil, and nuclear facilities, while the unavailability of gas and exploitable hydroelectric sites has nevertheless forced the industry to rely exclusively on these more expensive alternatives in expanding generation capacity. In some areas, substituting solar energy systems for electricity in selected uses, such as heating and air conditioning, appears to be an economical alternative to expanding generation capacity to satisfy growing energy demand.

Gas and electric utilities have begun to show considerable interest in exploiting solar technology. In some states, utilities have proposed that state regulatory commissions permit them to market solar energy as part of a combined energy package (Davis 1976). The issue facing regulators is how they should structure the emerging solar energy industry.

Solar energy presents three major problems to utilities and their regulators:

A portion of this chapter appeared in Federal Trade Commission, 1978. *The solar market: Proceedings.* Washington: FTC.

1. The technology of solar energy is not consistent with the natural-monopoly rationale for public utilities and thus threatens to erode the utility industries.
2. Solar rarely is an economical total substitute for conventional energy and, because of its dependence on weather conditions, may have little or no effect on peak demand and, therefore, the capacity requirements of conventional utilities.
3. For the most part, the existing methods of pricing gas and electricity provide the wrong incentives to utilities, utility customers, and commercial solar energy firms.

Each of these issues is examined in this chapter. The chapter first considers how the present structure of the utility sector is affected by, and will affect, the development of commercial solar energy systems and the likely consequences of alternative structures. The chapter then turns to an examination of the relationship of the pattern of energy demand and of pricing policies to the attractiveness of solar energy systems. The underlying theme of both sections is that an essential component of a rational public policy toward solar energy, or toward any "exotic" energy source, is a carefully planned restructuring of existing regulatory policies toward the energy sector that makes the pattern of incentives facing the users and suppliers of energy more consistent with efficient use of energy resources.

Solar Energy and Centralized Utilities

Although electricity and other distributed forms of energy can be produced at a central location by using solar energy, the most economical methods of using solar energy, at least for the foreseeable future, will convert solar radiation to usable energy at the point of use. This is in sharp contrast to the structure of utility firms. The essence of a retail energy utility is the transportation of energy from a central source to the point of use. Electrical energy is most efficiently generated in large facilities serving thousands of users. Natural gas is found in a relatively few scattered gas fields and transported through pipes to the homes and businesses that use it.

If solar energy were to capture a large proportion of the market from gas and electricity, it could make some of the existing distribution capacity obsolete. Moreover, gas utilities, especially, sell a fuel that appears to be in very limited long-run supply. Consequently, they face a slow corporate death as other energy forms replace the share of gas in the total energy market. Thus, on the one hand, solar energy systems do not appear to have the technical characteristics of public utilities, any more than do household appliances—each point of use can have its own, independent solar energy system without any cost sacrifice. On the other hand, if utilities (and especially gas utilities) do not capture a healthy share of the solar energy business, they may face slow extinction as the distribution systems of retail utilities are replaced by on-site solar collection systems.

The problem facing regulators is to determine the extent to which utilities will control the rate and pattern of development and use of solar energy technology. Obviously, a prime force in directing regulatory policy will be the generic difficulty government officials face in adopting policies that erode the welfare of any business: witness the problems of the Federal Communications Commission in permitting the demise of telegraphy or of Congress in letting rail passenger service disappear or in permitting the bankruptcy of Lockheed. Nevertheless, this chapter proceeds under the assumption that efficient operation of the energy sector is one of the desiderata of energy policy and examines alternative structural arrangements and pricing policies of electric and gas utilities in terms of their effects on the development of new energy systems such as on-site solar energy equipment.

Alternative Regulatory Policies

The alternatives available to the regulators fall into a few general categories. First, public utilities could be given exclusive monopoly franchises to provide solar energy systems to substitute for some of or all the other forms of energy used by their customers. Thus, gas companies would have exclusive rights to construct solar-assisted gas heating systems, such as Project SAGE, a system now under development to provide combined solar/gas hot-water heating (Davis and Bartera 1976). Of course, granting an exclusive monopoly to a utility for providing an integrated energy system incorporating solar equipment would be accompanied by conventional public-utility regulation of the prices and profits of the solar component of the system, just as the conventional component is now regulated. One approach is to set a regulated energy price that applies to all energy used by the customer whether from solar collectors or conventional sources. Then the gas company would, in effect, own the capital investment in the solar collection system and would set energy prices so as to return whatever rate of profit the regulators would permit on invested capital. Another approach is to allow customers to purchase solar equipment from the utility, with the company making separate charges for the capital investment (either a lump sum or an amortized installment plan) and for the amount of conventional energy consumed.

A second alternative is to deny utilities exclusive right to sell solar energy systems, but permit them to enter the solar energy business as part of their regulated public-utility activities. The utility would offer services as in the first case, except that customers could turn to nonutility firms to acquire the solar component of total energy service. Gas and electric utilities would offer regulated conventional service and regulated solar versions of the same service, but would face competition in the solar component. A customer desiring a solar energy system could either buy it from the utility or continue to buy straight gas or electric service from the utility but tack on a solar component that was purchased from a third party.

The third alternative is to allow utilities to sell solar systems through a separate, unregulated affiliate. Public utilities would face competition from nonutility organizations for solar equipment. Customers facing either gas curtailments or simply lower total costs from integrated energy systems would select solar equipment from among the unregulated competitors.

The fourth alternative is to prohibit utilities from selling on-site solar energy systems or the energy derived from them. Utilities could still explore the possibilities for using solar energy as a centralized energy source, either to generate electricity or to gasify hydrocarbon fuels, which would be tied to the utility distribution system, but selling or renting solar energy converters at the point of use would be denied them.

All these alternatives have precedents in the regulated utility sector. Until the late 1960s, telephone companies had exclusive control over the devices interconnected to the switched telephone system. A user seeking, say, to acquire an internal switching system that allowed phones in the same business to call one another without going through the local telephone exchange had to buy that system from the local telephone company. Just as solar energy can substitute for some uses of gas or electricity from a utility, so, too, do extension-to-extension calls in the same business substitute for use of the local telephone exchange. Even though there is no natural monopoly in internal switching systems, until the late 1960s, regulators nevertheless saw fit to extend the exclusive franchise of the telephone company to these and all other devices that could be attached to the telephone network.

Since the Carterfone decision by the Federal Communications Commission (*FCC Reports* 1968), telephone companies have retained the right to sell terminal devices, but they no longer have an exclusive monopoly. In principle, customers of a telephone company can purchase their own telephones from an unregulated entity or continue to accept instruments from the telephone company at a regulated rental rate that is included in the price of standard service. Other terminal devices can be rented from the telephone company at regulated tariffs or purchased from competitive suppliers.

Most utilities have business activities of the third type, that is, certain unregulated activities that are related to the regulated service. Some gas and electric companies sell home appliances, and some sell repair and maintenance services. Some telephone companies may either sell communications equipment at unregulated prices or include the use of such equipment as part of the company's service at regulated prices.

Finally, the fourth arrangement applies to one major utility, the American Telephone and Telegraph (AT & T) company, which is prohibited from engaging in unregulated business except, to a limited extent, in the sale of communications equipment. The 1956 Consent Decree that settled an antitrust complaint against the Bell System (*U.S.* v. *Western Electric and AT&T* 1956) prevents AT&T from entering such businesses as data processing. Computer time-sharing services, which rely on telephone lines to interconnect terminals and central processors, can be offered by telephone utilities not affiliated with Bell, but cannot be offered by Bell System affiliates (*FCC Reports* 1971).

Thus, history provides no consistent guide to the appropriate way to deal with emerging solar energy technology. Nevertheless, because similar regulatory problems have arisen in the past, the conceptual issues involving utility regulation that are pertinent to a decision on the structure of the solar energy business have been reasonably well formulated. Each potential market structure for a new technology, such as solar energy, that threatens an established regulated one, such as conventional retail energy, has a range of potential advantages and disadvantages. Because the importance of these advantages and disadvantages varies according to the particular circumstances of the case, the relevant task is to identify these issues and their likely importance in the case of solar energy development.

Arguments against Solar Ownership by Utilities

Public economic policy in the United States, at least since the late nineteenth century, begins with a presumption in favor of competition. The major exception to this principle has been the public-utilities sector. Around the turn of the century, most communities accepted the argument that public utilities were natural monopolies and conferred on these firms the right to seek monopoly status in return for subjecting their prices and profits to public control through a regulatory authority. In all but about forty cities across the country, a single firm now supplies electricity, gas, or telephone service in any given area, and in all but one state these activities are regulated by a state public-utilities commission.

Since solar energy technologies, at least for the near future, do not exhibit the kind of scale economies that may lead to natural monopoly in the utility sector, there is a presumption that they should be provided in competitive markets. Moreover, some additional problems can be anticipated from allowing energy utilities to provide solar technology.

If regulation is effective in keeping prices below the rates a monopolist would charge, it must necessarily create some perverse incentives which lead firms away from providing service at the lowest possible cost. For example, regulated energy utilities can earn profits only on investments in physical capital. They cannot earn profits on maintenance and installation activities or on the resale of energy. Consequently, a regulated utility has an incentive to engage in excessive substitution of capital investments for other productive resources (Averch and Johnson 1962). Moreover, because the rate-making process is based on historical data and consumes considerable time and resources in reaching a decision on rate increases, utilities have an incentive to avoid financial and technical risks. The effect of these incentives on solar energy is to lead utilities to invest in solar technology that is too durable, that is excessively efficient in converting sunlight to usable energy, and that requires inefficiently little maintenance. These strategies increase the capital used in exploiting the technology and insulate the firm from the possibility of future increases in noncapital costs. If permitted, this would lead to excessive costs and prices for solar energy and slow adoption of the technology.

Similar incentives are also in operation when utilities decide whether to invest in solar or other technologies, but their net effect is inconclusive. Solar technology appears relatively intensive in its use of capital, which is attractive to utilities; however, as a new technology, it also may appear relatively risky. Whether utilities would, on balance, pursue solar technology with excessive vigor or timidity probably is not possible to predict in advance of an actual market test. Nevertheless, utilities did enter the market for nuclear generation facilities enthusiastically in the mid-1960s. Nuclear, like solar, offered considerable uncertainties owing to its incomplete state of development and the opportunity to employ a capital-intensive technology.

Regulated utilities can use solar technology strategically as a means to create internal subsidies within their price structures and thereby to recapture some of the monopoly profits that regulation takes away as well as to foreclose competition in the solar energy business. For example, a joint solar/gas utility would have to work out a method to allocate its costs between solar-assisted and gas-only services. If it could effect an allocation that, in fact, attributed too much cost to gas, it would succeed in taking advantage of its monopoly in the gas business to subsidize its solar energy business. Normally, an unregulated firm would not find such a strategy attractive. But regulation provides the incentive to engage in this behavior because of the possibility that this strategy will enable the firm to capture more monopoly profits from its regulated gas business.

In principle, perfect regulation could prevent these problems, but in practice state regulatory authorities lack the resources and information to maintain perfect scrutiny of utility operations. Because the utility is always more expert than the regulator on the technical and economic conditions facing the firm, a technological advance that provides more flexibility in firm operations can be used strategically by the utility to work a better deal from the regulated market (Noll and Rivlin 1973).

Internal subsidization of the sort described above is not a farfetched, abstract notion. To the contrary, it is pervasive in the public-utilities sector (Posner 1971). The most obvious manifestation is rate averaging, in which utilities charge the same price for a particular service regardless of intercustomer variations in the cost of providing service. Utilities have been especially prone to internal subsidization to protect against competitive incursions into their markets. Examples include declining price as a function of total use by electric utilities, as a mechanism to encourage substitution of electric for gas appliances, and the pricing of certain communications services by AT&T at less than incremental cost after the Federal Communications Commission permitted the entry of some competitive communications firms into long-distance communications.

The fundamental point underlying the previous analysis is that a regulated utility has both the incentive and, to some degree, the ability to maintain energy service as a franchised monopoly after cost conditions no longer support the conclusion that monopoly is natural. Consequently, pricing, investment, and even research and development strategies will reflect this objective. While regulators, conscious of this problem, can attempt to offset it, the information and resource advantage of the utility, coupled with the incentive of the regulators to avoid cataclysmic service or financial failures by regulated firms, will inevitably lead to less than perfect regulation. This opens the door to strategic actions by regulated firms which increase their profits and secure their monopoly positions, but which do so at the expense of energy users.

Problems of Liability and Quality Control

One issue raised in connection with the choice of a role for utilities in solar energy pertains to the locus of responsibility for the quality of solar energy systems. Solar equipment is normally intended to be a long-term capital investment, yet in the early years of the industry purchasers of the equipment will have little information about the durability of competing systems. According to one study, the early history of solar energy in Florida saw the marketing of some solar heating equipment that was unreliable and subject to severe leakage problems (Booz, Allen, and Hamilton 1975). The standard approach to questions of product quality is to rely on brand reputations, voluntary trade association standards, warranties, and producer liability to generate adequate incentives for manufacturers to produce reliable equipment. Numerous household appliances—water heaters, stoves, and furnaces—are as capable, in principle, of causing severe damage as are solar heating and air-conditioning units; yet the incentives operating on manufacturers are sufficient that product quality in this area does not constitute a severe social problem. For example, as of mid-1977, the Consumer Product Safety Commission (CPSC), in its four-year lifetime, had yet to open a standard-setting proceeding for any major home appliance. The products under examination by the CPSC tended to be either products that are inherently dangerous and frequently associated with accidents (such as power lawn mowers, book matches, and bicycles) or products of industries in which the industry itself sought to have mandatory standards imposed on it, as was the case with swimming pool slides.

In the long run, solar equipment does not appear to present any special problems of product quality beyond those that arise in the home appliance industry or with other products that account for a major proportion of the budget of consumers. To the extent that product risk is a threat to consumers, one would expect the development of warranties, brand identification, and voluntary standards to cope relatively well with the problem.

Nevertheless, as the industry develops, both producers and consumers of solar equipment undoubtedly will pass through a learning period that will be accompanied by the temporary successful marketing of equipment that proves to be less reliable than originally anticipated. If utilities act as the gatekeeper to solar technology for their customers, two advantages might accrue: (1) experience with numerous consumers and producers would enable the utilities to learn the comparative strengths of equipment more rapidly, and (2) the permanence and stability of the utility would offer protection to consumers should the liability for damages from solar equipment be placed on the utility. Yet there are also disadvantages. Utilities are likely to be interested in greater product quality than are consumers, for reasons discussed above, and the risk aversion induced on utilities by the regulatory process is likely, in any event, to make them reluctant to assume such liability without exacting an excessive price to cover contingencies.

Government can play a role in promoting reliable solar equipment. Government could regulate performance standards for the industry, but because technology is evolving rapidly in solar equipment, standards could become a serious impediment to the maturation of the industry. Standards regulation, because of its procedural requirements, is inevitably a slow process that produces rules which are difficult to change and which, therefore, retard product innovation. Another possibility is for the government to monitor the development of the industry, publishing information on the performance of various types of equipment as experience develops and imposing some sort of truth-in-packaging requirement on equipment manufacturers. The attraction of this approach is that it deals directly with the essence of the problem, which is to increase the rate of diffusion of reliable product-quality information among customers of solar equipment.

The Case for Combined Solar Utilities

Several arguments have been advanced in support of granting utilities the right to offer solar energy systems as a regulated activity. These arguments generally boil down to an assertion that only utilities have the proper incentives to push for solar energy. The argument can take many forms, including relatively unenlightening comments about the rationality of consumers and potential entrants into the solar energy business. But at the heart of the issue is the perverse incentive provided to energy users and indirectly to those who would enter the solar energy business by the form of regulated prices for energy. In all but a few cases, the price structure for electricity and gas does not reflect the true marginal cost of service. Consequently, the customers of the utility do not face proper incentives to switch to solar energy. This problem is compounded by the fact that solar energy systems are rarely a complete substitute for conventional energy.

While the method of utility pricing differs from jurisdiction to jurisdiction, the procedure usually has been to set prices equal to the average cost of service. In most cases, utility customers are divided into groups according to their type of business or use of energy. Each group is then charged a price equal to the average cost of service for the group, with common costs, such as overhead and unallocatable capital facilities, meted out among groups according to total annual energy use. In some jurisdictions, each group faces a tiered price structure, with the price of energy use declining as use increases. [By far the most complete description and analysis of regulation is Kahn (1970).] In a few jurisdictions attempts have been made recently to move toward peak-load pricing of electricity, but for the most part prices are averaged over peak and off-peak periods.

Until the late 1960s both economies of scale and technological change worked to force down the costs of additional energy supply, especially in the electricity business. Since electricity prices were not sensitive to peak demands, there was a tendency to overbuild capacity, but this was partly offset by the lower average cost of new energy. In this milieu, quantity discounts for energy use had some economic rationale, albeit incomplete because of the absence of peak-load pricing, in that increments to energy usually did cost less than the average cost of the total supply. In the case of natural gas, economies of scale in distribution systems and the absence of a significant cost difference between off-peak and peak periods made the declining-block-rate pricing system quite rational as long as natural-gas field prices were not increasing.

Three events changed the situation (Joskow 1974). First, stricter environmental controls raised the real costs of electricity generation, and since more rigorous standards were applied to new generation facilities than to existing ones, costs were increased more for new generation plants than for old facilities. Second, inflation and higher interest rates caused the costs of energy inputs to begin to rise more rapidly than the rate of technical progress in the energy sector, so that dollar (if not real) capacity costs began to rise rather than fall, as they had for several decades before. Since most regulatory commissions calculate prices based on the depreciated original cost of capital facilities, rapid inflation served to increase the allowed costs of new service but not of service from old capacity. Third, the success of the Organization of Petroleum-Exporting Countries (OPEC) cartel since 1973 has raised the price of not only oil but also all fuels substantially more than the overall rate of inflation. But long-term contracts and domestic fuel regulation have served to give utilities access to some fuel at old prices. Hence the average cost of both gas and electricity has been held below the price that would reflect the costs of expanding energy output, which would be based on the price of new gas, new domestic oil, imported oil, or synthetic hydrocarbon fuels. In electricity, the inefficiencies arising from the differences between prices based on average cost and the true incremental cost of energy began to exacerbate, rather than partly offset, the inefficiencies arising from the failure to use peak-load pricing.

The problem is illustrated in figure 6-1, where the solid line MC_c represents the marginal cost of energy from conventional sources at various proportions of present output, and the dotted line represents the current average cost AC_c and price P_c. For simplicity of exposition, problems of peak versus off-peak costs are ignored for the present; all customers are regarded as facing the same costs, and the price elasticity of energy demand is presumed to be zero. However, the same qualitative conclusions hold for more realistic but more complicated cases. The dashed line represents a hypothetical marginal cost of solar energy (MC_s), where increasing amounts of solar energy as a proportion of total energy supply read from right to left along the horizontal axis of the diagram as reductions in the proportion of energy from conventional sources. To minimize total energy costs for society requires expanding solar energy to the point at which its marginal cost equals that of other energy—to point p in the figure, where $(100 - p)$ percent of the energy supply comes from solar technology.

Unfortunately, energy consumers have no incentive at the hypothetical current regulated price P_c to purchase solar equipment at an implicit energy price P_s. The consequence of their continuing to buy energy from conventional sources is that total societal energy costs are larger than they could be by the amount of the shaded area in figure 6-1.

The failure of electric utilities and their regulators to adopt peak-load

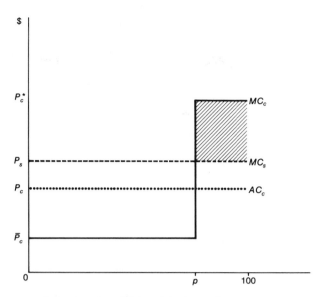

Proportion of Total Energy from Conventional Sources

Figure 6-1. Hypothetical Marginal and Average Costs of Solar and Conventional Energy in Relation to Conventional Proportion of Total Energy Supply

pricing produces a perverse incentive that favors solar energy. Solar energy systems are most efficient when sunlight is most intense, but the period of peak electricity demand may not occur during such periods. As a result, the installation of solar energy systems is likely to cause a greater reduction in electricity demand in off-peak than in on-peak periods. If a user faces electricity prices that are based on the average over peak and off-peak costs, the financial incentive to reduce off-peak electricity demand is as great as the incentive to cut back on peak load. For the utility a reduction in off-peak demand lowers its cost by much less than does a reduction in peak demand. Consequently, for some customers the financial incentive communicated by rate averaging may be sufficient to induce investment in solar energy systems that primarily substitute for off-peak electricity, even though the cost of the solar system exceeds the savings to the utility arising from the reduction in electricity demand because of the installation of the solar equipment.

Even though the utility price structure does not communicate appropriate cost signals to consumers, utilities nevertheless see the actual marginal costs of alternatives and can calculate the appropriate division of energy supply between conventional and solar sources. To return to the example in figure 6–1, a cost-minimizing utility would seek to induce customers to use solar technology until $(100 - p)$ percent of total energy was supplied from solar systems. Moreover, since solar technology is more capital-intensive than gas or electric supply systems, even if cost minimization is blunted by rate-of-return regulation, utilities are still likely to push the development of solar energy as long as it reduces total costs of energy. This is the basis for the argument that utilities should be given the right to sell solar equipment and/or the usable energy derived from it.

In the mid-1970s, utilities and their regulators began to consider revising the price structure for energy so that prices more closely reflected the true marginal costs of service. Several states have already adopted seasonally variable electricity prices to provide additional incentives to cut back electricity demand during the peak demand months of summer and winter, and widespread use of both rising block rates and time-of-day pricing may be in the offing (Mitchell, Manning, and Acton 1977).

A peak-load pricing system could eliminate most of the perverse incentive to adopt solar energy systems which cause an uneconomical reduction in off-peak electricity demand. Such schemes can not work perfectly unless quite elaborate metering and pricing systems are adopted so that electricity prices can vary over very short time intervals. Since the cost of implementing peak-load pricing increases with the complexity of the system, utilities and regulators are not likely to find a very elaborate scheme worthwhile; however, substantial savings in total energy demand and costs apparently are possible with relatively simple adjustments in the pricing structure. Apparently, the best route to solving the problem of excessive investment in solar equipment that leads to uneconomical substitution for off-peak demand is to adopt a more rational price structure for electricity.

Unfortunately, the problems associated with the rising costs of new energy resources cannot be solved within the context of rate-of-return regulation by reliance on prices based on marginal costs. Once again, the following analysis will abstract from the problems of peaking, although the analysis applies to this more complicated pricing problem as well. This analysis can be thought of qualitatively as applying to the price structure prevailing in each period, although a complete analysis would have to account for interdependencies among periods in costs, capacity, and demand. In figure 6–1 marginal-cost pricing would dictate a price of conventional energy equal to the topmost portion of MC_c, shown as P_c^* in the diagram. Eventually, prices set at P_c^* would induce solar energy development to $(100 - p)$ percent of total energy, at which time conventional sources would have a lower marginal cost. If we were to assume that the true conventional marginal-cost curve is not, as shown, discontinuous at p, but instead is merely rising in that region and passes solar energy marginal costs at p, then the long-run price for all types of energy would converge to P_s. Unfortunately, such pricing is inconsistent with much of the purpose of utility regulation. In the long run, utility prices equal to P_s would generate excess profits for the utility as measured by the area between the P_s line and the MC_c curve to the left of point p. And, since energy-using equipment is normally a rather long-lived capital asset, several years would transpire before solar energy accounted for $(100 - p)$ percent of total energy. Meanwhile, utilities would, at price P_c^*, earn even larger excess profits, equal to the area between P_c^* and MC_c to the left of point p.

The normal response to an incompatibility between pricing at marginal cost and limiting utilities to a competitive rate of return is to invoke a multipart price structure. Each consumer is asked to pay a low price for the first few units consumed that is based on the low cost of these units and then to face ever-increasing prices for additional consumption, with these prices reflecting even higher costs for newer sources of energy. In the simplified case shown in figure 6-1, consumers would face a price $\bar{P}_c$ for the first p percent of their anticipated consumption and then a price P_c^* for proportions above p of total energy use devoted to conventional sources.

Figure 6-2 represents a more complicated, and more realistic, hypothetical example of the utility pricing problem. Like figure 6-1, figure 6-2 abstracts from peaking problems, demand elasticities, and differences among energy users in the demand for energy or the cost of solar equipment. Unlike figure 6-1, the situation shown in figure 6-2 depicts the marginal costs of providing energy to a particular user for various proportions of that customer's total energy demand that is accounted for by conventional (MC_c reading left to right) and solar (MC_s reading right to left) sources. In figure 6-2, the solar energy cost curve is constructed to illustrate crudely the actual situation for a particular point of energy use: scale economies in solar energy systems for the first few percentage points of energy accounted for by solar and then rising costs as the solar proportion rises. The conventional-source cost curve (MC_c) assumes that all energy will reflect the high costs of new conventional sources, which enter the calculation of P_c^*. In the case shown, the minimum-cost energy system is to devote q percent of energy use to conventional sources and the rest to solar.

The diagram is constructed to illustrate a problem with the use of marginal-cost pricing to induce solar energy conversion. In figure 6-2, point p represents the same point as in figure 6-1, that is, the proportion of society's total energy demand that ought to be devoted to conventional energy, based on the lower costs of old energy sources. The dotted line in figure 6-2 depicts the price of solar energy if $(100 - q)$ percent of a customer's energy is from solar technology. The dashed line is the two-part price structure shown in figure 6-1 that prices conventional energy beyond p percent demand at the marginal cost of new supplies but conventional energy for

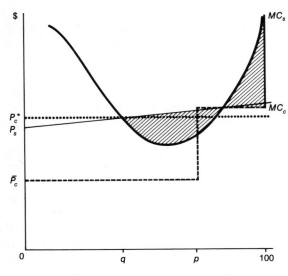

Figure 6–2. Hypothetical Marginal Costs and Prices for a Representative Customer for Solar and Conventional Energy in Relation to Customer's Proportion of Energy Accounted for by Conventional Sources

the first p percent of the lower average cost of old supplies.

As the diagram is drawn, the customer facing the high marginal cost of new energy over all ranges of energy use has an incentive to shift to $(100 - q)$ percent solar because the cost savings in the leftmost shaded area offset the cost disadvantages of solar in the rightmost shaded area. But if the price structure represented by the dashed line is imposed, letting customers face the appropriate price incentives for the last units of energy consumed but preventing the utility from earning excess profits, consumers will not use any solar energy. The reason is that at p percent conventional energy, solar prices must be higher than P_s. In particular, they must recover the high costs of the first few units of solar energy without benefit of the additional scale economies between p and q. As shown, the losses in the rightmost shaded area will offset the potential cost savings in that portion of the leftmost shaded area which lies to the right of point p. In other words, marginal-cost pricing of the latter sort, while providing proper incentives at the margin, still fails the total conditions for efficient energy selection, given the particular form of the cost curves assumed in figure 6–2.

The underlying problem depicted in figure 6–2 is that the optimal proportion of solar energy for a particular customer is not equal to the optimal proportion for society. Hence, the market signals conveyed by a multipart pricing structure based on marginal costs but returning no excess profits to the utility provide inappropriate incentives. In particular, the market does not induce a relatively small (smaller than p) percentage of energy customers to demand more solar energy (q percent of their total use) than is optimal as a societywide average. Consequently, no universal price structure can be constructed which by itself provides appropriate incentives to customers to develop solar energy and holds utility monopolies to a competitive rate of profit.

Some Solutions to the Policy Problem

If both the utility and the solar energy industries were perfectly competitive, an optimal proportion of solar energy would eventually develop. The reason is that competition would not allow price differentials to persist between old and new sources of conventional energy. Holders of old sources, such as old gas and oil wells or generation facilities constructed before the recent spate of inflation, would experience capital gains that valued the sources under their control at their opportunity costs in terms of new energy sources. All energy use would then be priced at the marginal cost of new sources, and utilities would still earn a competitive return—but on assets whose values reflected these capital gains. Of course, these capital gains would be equivalent in value to the excess of revenues over the marginal costs (based on original costs) shown in figure 6–1. The incompatibility between efficient adoption of solar technology and energy pricing policy arises because the latter attempts to prevent not only monopoly profits, but also the windfall gains a competitive industry would receive if old investments appreciated in value owing to an increase in the cost of new investments.

The preceding argument does not support the conclusion that a tiered rate structure is undesirable. If energy customers differ according to the attractiveness of solar energy to them, some justifiable switch to solar technology will be induced by making the price of the last units of conventional energy reflect their social costs. Moreover, if the assumption is relaxed that the demand for energy is perfectly inelastic, some additional gains in energy conservation will result from the kind of multipart price structure discussed here. What the preceding analysis does show is that this multipart price structure is likely to be insufficient to generate a socially efficient extent of solar development.

Granting utilities the right to sell solar equipment does not completely eliminate the problems inherent in the regulated price structure. In addition, some provision must be made for deciding which customers are forced to convert to solar systems. In the situation depicted in figure 6-2, some energy users would have to be required to acquire $(100 - q)$ percent of their energy from solar equipment despite the fact that they would be unwilling to do so unless their total energy expenditures were substantially higher than is the case under the existing rate structure for conventional energy.

Thus far regulators have considered two approaches to this problem. One is to allocate rights to low-cost conventional energy to particular groups of customers, as has been done with respect to natural gas. The second is to adopt a price scheme that recovers from all customers the average cost of providing solar energy to some and only conventional energy to others.

The first method would single out some energy customers to face conventional energy prices based exclusively on new energy sources, as reflected in MC_c in figure 6-2, for all energy consumed beyond q percent of total energy consumption. The remaining customers would face only the price based on old sources, $\bar{P}_c$. If the sizes of the two groups were appropriately picked, the end result would be the appropriate one: $(100 - p)$ percent of energy use from solar sources. In essence, this method transfers the windfall gains that would have accrued to owners of old conventional sources to customers, with most of the gain going to those customers in the high-priority class who face only the low energy price for all units of energy consumed.

The second method divides the same windfall gains to old energy sources equally among all consumers. Each customer would face a price structure based on the costs of old energy for the first p units of energy consumed and on the costs of solar for the remaining units. In figure 6-1 the price structure is represented by price $\bar{P}_c$ up to p percent of energy use and P_s afterward. Since this would not induce any conversion to solar by customers, the utility could then simply install solar energy systems in some locations, each of which would produce $(100 - q)$ percent of the energy on the site by solar technology but at no change in energy costs to the particular customer who received the solar converter. Of course, the utility would have to be empowered with the right to require that the appropriate proportion of customers adopt solar-assisted systems, subject to some sort of regulatory and/or judicial review with regard to forced conversions. Regulators could guide the coercive conversion program by establishing priorities by user class, as in the first method.

The first method does not require that utilities be engaged in the solar energy business. Customers granted limited access to conventional energy sources have adequate incentives to make optimal choices among alternative strategies: purchasing energy from new sources (such as solar) and engaging in conservation. Customers with unlimited access to conventional energy at the low historical cost have no incentive to convert to other sources. If the size of the latter group is correctly selected, this will not lead to a socially inefficient amount of solar energy conversion. But this method will not, in general, provide appropriate incentives to invest in energy-conserving activities, such as insulation. Even if all customers, including the favored group, face a marginal energy price that reflects true marginal energy costs, the same failure of the price system discussed above in the context of solar conversion can occur. That is, the marginal price of energy may prevail over too narrow a range of energy use to induce switching to a conservation technique that exhibits scale economies at each site of use.

The second method, as described, does require utilities to enter the solar energy business, for only they will have an incentive to make appropriate conversions to solar systems. Moreover, this alternative requires considerably more regulatory supervision than the first, in part to limit inefficient development in solar technology owing to the perverse incentives presented to firms by price regulation and in part to provide due-process protection to individuals who are coerced into permitting the installation of solar energy equipment on their property. Finally, as with the first method, this approach will not, in general, provide adequate incentives for energy conservation.

By casting both methods in the form of systems to allocate property rights to old sources of energy, a third method emerges that would achieve efficient conversion to solar or other exotic sources of energy and conservation without requiring that utilities be allowed to enter the solar business or that coercion be used to induce solar development. In addition, it would achieve another goal of the second method, which is to spread the windfall gains that would accrue to owners of old energy resources relatively equally among energy consumers. All these objectives could be accomplished by the following policy. First, energy prices would be allowed to rise to the point that would prevail if all windfall gains were actually captured by owners of old energy sources (to price P_c^* in the figures). Second, coupons would be printed that entitled the holder to purchase a unit of energy at the price based on the original cost of old energy resources ($\bar{P}_c$ in the figures). The number of coupons printed would, in total, create rights to purchase all the inexpensive energy that was available at its true cost. Each customer would be given some number of coupons based on the equity objectives of the government. Any mechanism of distributing the coupons, including a proportion of historical use or an equal per capita amount, will work as long as one's current energy use has no effect on the future distribution of coupons. Finally, energy users, utilities, and anyone else would be permitted to buy and sell coupons.

The market price for coupons would be established quickly at the difference between old and new energy prices, and some energy users would perceive that by selling coupons at this price and installing solar energy they could come out ahead. Moreover, if utilities were permitted to buy coupons, in addition to being required to accept them in lieu of utility bills, at whatever price was acceptable to consumers, the rapid establishment of a coupon market would be ensured. If utilities could buy coupons at a price somewhat below the difference in old and new energy costs in return for providing a convenient coupon market for consumers, they could also come out ahead since a coupon purchased by a utility would represent one less unit of energy that had to be sold at the old price.

The preceding mechanism is a natural extension of the method used in the 1960s to allocate oil import quotas and the entitlements program currently used by the Federal Energy Administration to prevent owners of old petroleum reserves from capturing windfall gains owing to the increase in the world price of oil that occurred in 1973. In essence, the coupon proposal establishes entitlements among consumers to cheap energy sources. In addition, it has an advantage not shared by the entitlement program in that it allows energy prices to reflect marginal costs. Because entitlements in the oil refining business are based on current rather than historical energy use, the latter program also causes the price of oil to refiners to equal average rather than marginal cost, thereby encouraging uneconomical use of petroleum (Montgomery 1977).

Creating tradeable rights to cheap energy allows the government to allocate the benefits accruing from cheap energy sources in any way it sees fit simply by appropriate choice of the initial distribution of the coupons. Once the coupons have been distributed, prices of energy can be set equal to marginal cost without fear of windfall gains accruing to producers. Instead, the windfall will accrue to coupon holders. Moreover, the system creates the proper incentives for individual energy users to switch to new energy sources and to engage in energy conservation. Consequently, it eliminates the argument in favor of utility ownership of solar energy resources.

Conclusions

Because of the unavoidable imperfections in regulation, a decentralized, competitive solution to the problem of solar energy development has great merit. The use of tradeable property rights and peak-load pricing allows competitive, unregulated solar energy development and provides appropriate incentives to consumers in selecting energy sources.

Whether utilities should be permitted into the on-site solar energy business at all remains an open question: there are dangers owing to imperfections in regulation, but there is promise in that gas utilities, in particular, are currently at the forefront of at least some types of solar equipment and possess an especially strong incentive (namely, long-run corporate survival) to make the technology work. Perhaps the best alternative is to allow utilities to enter the business, but to require that their solar energy activities be handled through separate, unregulated affiliates and that solar and regulated utility services be sold and priced independently. This would enable a solar energy customer to select from competitive suppliers while facing a price for conventional service that would not depend on the source of solar equipment. A utility might still decide to operate a solar energy affiliate at a loss, but at least such behavior could be attacked relatively easily through antitrust or regulatory action owing to the separation of solar and other business activities and, therefore, the relative ease with which cash flows could be observed.

Utilities would still have something of an advantage in selling solar technology, in part because of their experience in retailing energy and in part because of the relatively low costs of capital they face as a result of their protected, regulated status in mainline activities. Although the magnitude of these advantages is surely not known, it is unlikely to be sufficiently important that it would preclude entry into the solar equipment business by an efficient potential supplier.

The preceding analysis indicates that the current problem of promoting efficient energy use while preventing windfall gains to owners of old energy sources has a relatively simple resolution. The heart of the problem lies in rationalizing the price structure for energy and allocating the rapid increase in the value of cheap energy resources that has occurred because of increases in the costs of new energy sources. A system of tradeable energy coupons can solve the problem of windfall gains without placing limits on market prices, and a cost-sensitive price structure, including time-of-day variations in prices, can generate the proper incentives for choosing among alternative energy sources. Consequently, the equity problem can be solved with one instrument, and the efficiency problem can be solved in the market. This eliminates the need for expanding the domain of utility regulation into the solar energy field by either state or federal regulatory authorities. It does require that utilities not be permitted to start a regulated solar energy business, but there appears to be no strong argument for keeping them out of the unregulated solar energy business as long as they do so through completely independent corporate affiliates. A program along these lines seems capable of serving the twin objectives of efficient use of solar and other energy technologies and equitable distribution of the societal costs of rising energy prices.

7 The Solar-Utility Interface: A Capsule Appraisal

The existing centralized infrastructure of the utility industry plays a prominent role in the development of decentralized energy sources. This role is described by the progression of thoughts highlighted in this book. The consumer of decentralized applications makes his/her own choice of technology based on his/her response to price signals from competing energy sources. The rates or tariffs for backup electricity will determine the mode of choice for sizing and operation of solar applications and energy conservation. The price signals presently given by the utility industry are highly volatile and not representative of social costs. Their departure from social costs disturbs appropriate consumer responses and thus consumers' choice. Approaches to reaching societal optima are impeded by this pervasive condition. Regulatory and legislative response to this imperfect situation is vital.

Tax credits, a partially remedial solution, must be designed to promote the basic conditions of allocative efficiency in the attainment of policy objectives. Tax credits will affect consumer choice on the timing of the adoption decision and the sizing and type of the system to be adopted.

The electric-utility sector can provide stimulus to the adoption of decentralized energy sources through incentives for load management. The utility industry already recognized the need for those conditions, but it must be provided with analytical tools and data for massive implementation. The utilities can have the opportunity to use decentralized sources to lessen the need for more expensive capacity additions. Reluctance on the part of the utility sector to participate more actively in such programs may be a result in part of utility history, which is supply-oriented. The perception of the adverse impact of decentralized technologies on utility finances needs to be addressed. The varying nature of the impact of different technologies is a dynamic element which has to be incorporated into utility planning.

This book discusses the proper methodology for the individual homeowner in choosing among these many options for space conditioning and heating water. The methodology includes the important perspective of the utility industries that may compete with these new technologies and may supply auxiliary power. The methodology also provides a guide to government policymakers that leads to strategies for maximizing societal benefits.

Several recent reports have suggested that mechanical solar heating systems with collector, transport, and storage subsystems—*active solar sys-*

tems—can economically replace, in part, presently utilized space-conditioning technologies. The reports have suggested that active solar energy can reduce the national problem of high dependence on foreign fossil fuels. As a result of these studies, federal policy has been developed to promote the commercialization of active solar energy systems.

Other techniques exist that may also lower consumer energy bills and dependence on foreign fuel supplies, including improved building design, passive solar energy, and energy conservation. The feasibility studies for active solar energy have not considered these other technologies. Because of this, the studies cannot draw conclusions as to best overall investments among these technologies for homeowners or society. These studies also do not include the costs to the electric utility that is required to provide auxiliary energy. More importantly, these studies do not clearly distinguish differences that may exist between the individual consumer and society as to optimal investment strategies. The development of a more systematic methodology will reveal the flaws in these early reports and suggest more sound public-policy direction.

This book stresses the importance of a "systems" framework in the optimization of residential space conditioning. The home is a complex system; the occupant, the building shell, the internal living space, the HVAC equipment, and the external environment are all significant factors. Earlier studies have concentrated on only one of these aspects in an attempt to improve space-conditioning costs. Specifically, solar energy studies have assumed that the building heating and cooling load is fixed with respect to other aspects of the home and that maximization of benefits entails only adjustments in solar energy equipment sizing.

Actual optimization of consumer cost-effectiveness is achieved only through consideration of every factor affecting building load. Building loads can also be reduced through utilization of larger insulation levels, alterations in consumer use patterns, and improvements in building design. Many architectural practices long regarded for their ability to reduce heating and cooling loads have become increasingly underutilized because of the low costs of HVAC equipment. Proper consideration of these practices has not been included in these feasibility studies. One technique, involving the direct utilization of solar radiation integrated into the design of the building structure—*passive solar energy*—has demonstrated its effectiveness.

The homeowner has the eventual responsibility for choosing among these techniques. Ideally, the optimal choice would be the set of options among all conceivable combinations that produces the lowest total cost. The consumer is concerned with having low heating bills, but he also must weigh these *variable* costs against *fixed* investments such as solar equipment, energy conservation, and the structure itself. Despite ultimate savings resulting from lower variable costs, large initial expenditures in fixed costs constrain consumer choices. To a large extent the homeowner is limited by barriers in the building industry that resist large capital expenditures and new practices which may slow construction or increase costs. Lack of perfect knowledge about the tradeoffs between investments, and particularly about future prices of fuel, further affects consumer choice. This leaves a large degree of sensitivity in the eventual specification of consumer cost-effectiveness.

A second problem that exists among solar feasibility studies is the failure to identify differences that may exist between consumer cost-effectiveness and societal optima. Many reports have identified the set of cost-effective options for the consumer as being the set which best reflects benefits to national energy policy. The existence of inefficiencies within the marketplace causes deviations to exist between these two perspectives. A closer examination of the entire energy network is warranted if better government policy is to be derived from solar feasibility studies.

The social costs of dependence on foreign energy supplies, deficits in balance of trade, and environmental spill-overs are not reflected in the present cost of fossil fuels. In addition, government subsidies to promote these fuels have maintained prices that are lower than the true *marginal cost,* or replacement cost, of production. [For a discussion of the preferential tax treatment of public utilities' investment as opposed to decentralized solar owned by homeowners, see Chapman (1979).] This underpricing of traditional energy sources not only causes new technologies to be less competitive, but also shifts solutions of maximum national economic efficiency away from optimal consumer cost-effectiveness.

The role of the price of alternative fuels is even more important with respect to solar buildings which not only compete with traditional energy sources but may also utilize these sources for auxiliary supply. Most studies of solar feasibility have not fully regarded the implications of this relationship. These studies have assumed that calculations of average monthly auxiliary consumption are sufficient to calculate costs to the consumer and to society. In fact, temporal variations in cost of supply exist within public utilities. This is particularly the case in electric utilities where variations in load by time of day and season may result in cost differences significant enough to warrant differential rates.

Variations in the cost of electricity supply result between residences because of variations in the amount and temporal nature of electricity consumption. Peak-load demands for electricity have higher costs of electricity supply because they require the electric utility to utilize less efficient generation plants and to ultimately increase the generation capacity. Solar buildings may impact on electric utilities by lowering annual consumption without a commensurate reduction in generation capacity. These buildings may also have the potential to aid in electric utility *load management* by successfully lowering demands at times of utility peak loads. For utilities in California, Colorado, New Mexico, Massachusetts, and Texas that were examined in this book, it was shown that solar consumers can hardly be distinguished from nonsolar customers who are energy conservers in terms of load pattern. It appears that solar consumers, ceteris paribus with regard to the quality of service, may not constitute a separate rate class. Given these conditions, unless proved otherwise in particular circumstances, electricity tariffs should be based on long-run marginal-cost principles.

The present price of electricity is therefore an inadequate measure of cost of supply of electricity. A detailed examination of the cost of service provides a better indication of costs of auxiliary to the solar building. In addition, new tariff designs, which better reflect the cost of electrical service, will affect the relative cost-effectiveness of space-conditioning technologies.

To the extent that these omissions existed in previous studies, the development of a more comprehensive methodology was a major focus of this book. The goal of the methodology was to establish a "total system" framework which considered the perspectives of the individual, the electric utility, and society. The methodology developed allows the comparison of individual building components to determine the relative contribution of each component to lower space-conditioning requirements and utility auxiliary requirements. The methodology relies on the use of a building simulation model, in this case a modified version of TRNSYS. The use of simulation in the calculation of building loads is required despite some questions as to its validity because of the lack of usable performance data from actual buildings.

Significant advantages are apparent in the methodology utilized in this book versus those of previous solar feasibility studies. The analysis is calculated on an hourly rather than monthly basis. The TRNSYS simulation allows building load to be treated dynamically rather than statically, which better accounts for the lag effects of heat transfer within the building envelope. The design procedure permits more detail in building design, accounting for effects of changes in design, insulation levels, and building materials. Finally, special alterations to the original TRNSYS model enable the modeling of one passive solar technique—direct solar gain.

The integration of the individual building load into the utility load allows an assessment of impact on utility load, generation requirements, and finances. This capability improves the public-policy process by calculating a truer assessment of societal benefits. The breakdown of individual building components on a microlevel can be better extended to broader market-penetration projects.

In order to demonstrate the appropriateness of the methodology, a detailed case study was performed. The selected location, Colorado Springs, Colorado, represents a growth area with a long heating season and a strong interest in solar energy development. The electric utility is the municipal utility serving the Colorado Springs vicinity. The utility is characterized by electricity production costs and rates lower than those of the average U.S. utility.

The book begins with the designation of a base building onto which the feasibility of alternative techniques is tested. The book examines, for new construction only, the cost-effectiveness of each of the building practices, including insulation levels, different building materials, active solar-heating systems, active hot-water systems, fenestration techniques, passive direct gain, off-peak thermal energy storage, and conventional electric heating and cooling. The analyses are performed for several different future scenarios and tariff designs.

The results for the first-year analysis reveal that the least-cost alternative of those options examined is the all-electric base building, a 2 × 4 in wood-frame, fully insulated stud wall with 6–in attic insulation. Alternatives that reduce electricity consumption such as solar heating or energy conservation were not found to be economical under this scenario. Rising fuel costs, however, result in benefits to the consumer from energy-saving purchases over the lifetime of the investment. By using a life-cycle cost accounting system, these future savings are better reflected, although the individual cannot always fully appreciate these future benefits.

The results indicate that energy conservation directed to a 2 × 6 in fully insulated stud wall and 12–in attic insulation is warranted when future benefits are considered. The more future-oriented the scenario, the more cost-effective combined investments in masonry walls, large south-facing window areas, and mechanical insulating shutters become. These latter techniques have been found to lower heat loss, increase useful direct heat gain, lower cooling loads, and store thermal energy to slow internal room-temperature changes.

Investments in active solar energy systems do not appear to be economical compared to either passive solar energy or energy conservation even under the most favorable economic assumptions for Colorado Springs. Another significant finding is the poor economic performance of buildings without wall insulation. The life-cycle cost of a building without wall insulation is twice the cost of a building with wall insulation—in one example, $40,000 versus $20,000 despite an initial cost difference of less than $400.

The importance of a systematic methodology can be seen by comparison of two buildings with and without wall insulation. A 35-m² collector/storage system on each of the two buildings provides over 4,000 kWh more useful energy in the unisulated-wall building. The cost-effectiveness of that solar system is more pronounced in the uninsulated-wall building. However, solar investment in the unisulated building is a poor choice in comparison to an investment in wall insulation.

The cost-effectiveness analysis as expressed above does not provide a complete analysis of tradeoffs or impacts. There is a need to determine the impact on the electric utility that provides backup electricity. A significant finding of this book is the variation in impacts on the electric utility of different building components and designs. Buildings with solar energy systems were found to lower the peak demand of the winter-peaking Colorado Springs utility relative to similar buildings with all-electric heating and cooling equipment. At the same time, however, reductions in the annual consumption of electricity and the *individual building load factor* were far greater for these solar buildings as opposed to the all-electric counterparts. Solar-heated buildings had individual load factors below 0.10, representing a potential problem for electric-utility capital and generation capabilities.

Other building investments were found to lower peak demands at a much greater proportion of total kilowatt hour consumption. Individual building load factors were improved by the use of the integrated passive system design modeled in this book. The use of thermal shutters in large amounts of window expanse was found to improve individual load factors. In most cases, energy conservation and building thermal mass improved or maintained the individual load factor in comparison to the all-electric base building.

Solar-heated buildings have been found to cause large deficits in electric-utility revenue versus cost of service for the heating and cooling portion. The solar buildings were found to have large capacity-related (kilowatt) costs that were not adequately recovered through present tariff designs which average these capacity-related charges over a larger per kilowatt hour basis. Concerns of this kind by utilities have created interest in different tariff designs for solar-heated buildings. The impact of these new rates will ultimately affect consumer energy investment.

The benefits of energy investment to the consumer must be weighed against the impacts of that investment elsewhere throughout society. The electric-utility interface is one of those impacts which must be addressed. Equally important are the positive benefits that accrue from reduced consumption of fossil fuels. Unfortunately, the quantification of these costs is difficult and involves considerable value judgment. This book has demonstrated that the benefits of wall insulation are clear even without any inclusion of unquantifiable social benefits. For $10,000 spent in new housing for the insulation of uninsulated stud walls, almost $140,000 worth of actual incurred benefit is returned to society in fuel savings and generation capacity forestalled, over a 20–yr life-cycle scenario. Benefits over the 20–yr study period are also realized in 110 kW of capacity forestalled and 7.7 million kWh of fuel saved.

The benefits to society of wall insulation appear to be so great in Colorado Springs that it is difficult to contemplate a situation where new housing should be permitted without obeying high standards. This result appears to be substantiated in other areas, including California, Massachusetts, New Mexico, and Texas. While this assessment is strictly for new construction where cost estimates for wall insulation are $300 to $400, retrofit practices can be justified if costs are not more than ten times this cost. It also appears that more energy conservation, such as the construction of 2 × 6 in stud walls and/or 6–in additional attic insulation, and passive solar energy produce in Colorado Springs societal benefits greater than are produced by active solar energy. These additional investments in energy conservation return greater societal benefits in economic efficiency, capacity forestalled, and energy saved than do equivalent investments in active solar energy. These findings reveal what is perhaps a major discrepancy between the societal optimal investment strategy and present energy policy.

Although controversy exists as to whether utilities should offer financing arrangements for solar and energy-conservation consumers, it has been shown that such arrangements are cost-competitive with additions to utility capacity. It is largely a problem of how one specifies the financing arrangements as well as the fraction of finance that should go to solar as opposed to energy conservation.

It should not be without consequence if the majority of utilities do not take the same aggressive stance for energy conservation and solar energy as they did for expanding markets for power in the 1960s. If it does become apparent that utilities are reluctant to take an active hand, other institutions and policy directives could be created to fill the hiatus. It is not inconceivable that "solar utilities" could be formed on cooperative or other bases. There is a precedent formed by the Municipal Water Department of Santa Clara, California, and the Tennessee Valley Authority in selling and installing solar applications. As this book goes to press, Public Service Electric and Gas Company of New Jersey, and Wisconsin Power and Light Co. are launching solar hot-water heater programs for residential customers. In addition, California regulators have just ordered utilities to finance 175,000 solar water heaters over the next three years. It is hoped that the methodology provided here will aid policymakers in assessing the merits of such a program on an individual utility basis, and will provide federal and state policymakers with the tools for assessing the impact of public-policy options on such proposals.

References

Abrash, M.; Wirtshafter, R.; Sullivan, P.; and Kohler, J. 1978. Modeling passive buildings using TRNSYS. *Proceedings of 2nd Passive Solar Conference,* Philadelphia, Pa.

Aerospace Corporation. 1978. *Solar heating and cooling of buildings (SHACOB) requirements and impact analysis.* Palo Alto, Calif.: Electric Power Research.

Aigner, D. 1977. Models and methods for analyzing residential TOD experimental data. Unpublished draft.

———. 1979. "New directions in load forecasting with emphasis on time-of-use analysis." Unpublished draft.

Amaroli, G. California Public Utilities Commission, Personal communication.

American Institute of Architects (AIA). 1976. *Early use of solar energy in buildings: A study of barriers and incentives to widespread use of solar heating and cooling systems.* AIA, Summary Report to National Science Foundation, APR 75-18339.

Anderson, B. 1976. *The solar home book.* Harrisville, N.H.: Cheshire Books.

———. 1977. *Solar energy: Fundamentals of building design.* New York: McGraw-Hill.

Anderson, K.P. 1972. *Residential demand for electricity: Econometric estimates for California and the United States.* The Rand Corporation (R-905-NSF).

———. 1973. *Residential energy use: an econometric analysis.* The Rand Corporation (R-1297-NSF).

Arthur D. Little, Inc. 1975. *An impact assessment of ASHRAE standard 90-75, energy conservation in new building design.* Cambridge, Mass.: A.D. Little, Inc.

———. 1977a. *Solar heating and cooling of buildings commercialization report, Part B.* Cambridge, Mass.: A.D. Little, Inc.

———. 1977b. *System definition study—phase 1: Individual load center, solar heating and cooling residential project.* Palo Alto, Calif.: Electric Power Research Institute.

———. 1978. *EPRI methodology for preferred solar systems (EMPSS) computer program documentation.* Palo Alto, Calif.: Electric Power Research Institute (EPRI).

———. 1979. *Solar heating and cooling simulation programs: Assessment and evaluation.* Palo Alto, Calif.: EPRI.

Asbury, J.G.; Maslowski, C.; and Mueller, R.O. 1979. Report. *Science* 206:206-208.

239

Asbury, J.G. and Mueller, R.O. 1977. Solar energy and electric utilities: Should they be interfaced? *Science* 195:445–450.

ASHRAE. 1974. *Handbook of applications.* New York: American Society of Heating, Refrigerating, and Air Conditioning Engineers (ASHRAE).

ASHRAE. 1975a. *Energy conservation in new building design, ASHRAE 90–75.* New York: ASHRAE.

ASHRAE. 1975b. *Procedure for determining heating and cooling loads for computerizing energy calculations.* New York: ASHRAE Task Force on Energy Requirements.

ASHRAE. 1977. *Handbook on fundamentals.* New York: ASHRAE.

ASHRAE. 1978. *Handbook on systems.* New York: ASHRAE.

Averch, H., and Johnson, L.L. 1962. Behavior of the firm under regulatory constraint. *American Economic Review* 52:1052–1069.

Baer, S. 1975. *Sunspots.* Albuquerque, N.M.: Zomeworks Corporation.

Balcomb, J.D. 1976. Summary of the passive solar heating and cooling conference. In *Passive solar heating and cooling.* Washington: Energy Research and Development Administration.

Balcomb, J.D. and Hedstrom, J. 1976. A simplified method for sizing a solar collector array for space heating. Paper LA–UR–76–160, Los Alamos Scientific Laboratory.

Balcomb, J., Hedstrom, J.C.; and McFarland, R.D. 1977. Passive solar heating of buildings. Los Alamos, N.M. SAND–77–1204, Sandia Laboratory.

Balcomb, J.D. and McFarland, R.D. 1978. A simple empirical method for estimating the performance of a passive solar heated building of the thermal storage wall type. *Proceedings of the 2nd National Passive Solar Conference,* Philadelphia, Pa., ed. D. Prowler. Killeen, Texas: American Section, International Solar Energy Society.

Barkley, P.W., and Seckler, D.W. 1972. *Economic growth and environmental decay: The solution becomes the problem.* New York: Harcourt Brace Jovanovich.

Barnett, H.J., and Morse, C. 1968. *Scarcity and growth: The economics of natural resources availability.* Published for Resources for the Future, Inc. Baltimore, Md.: Johns Hopkins Press.

Baughman, M.L., and Joskow, P.L. 1974. Interfuel substitution in the consumption of energy in the United States. Draft, Massachusetts Institute of Technology, Cambridge.

Beckman, W.A.; Klein, S.A.; and Duffie, J.A. 1977. *Solar heating design by the F-Chart method.* New York: Wiley.

Ben-David, S., Schultz, W. 1977. Near term prospects for solar energy: An economic analysis. *Natural Resources Journal* 17:169–207.

Bennington, G., 1976. *An economic analysis of solar and space heating.* McLean, Va.: The Mitre Corporation, ERDA Contract E (49–1)–3764.

Bennington, G.; Cunto, P.; Miller, G.; Rebibo, K.; and Spewak, P. 1978. *Solar energy: A comparative analysis to the year 2020.* McLean, Va.: The Mitre Corp., Metrek Division. ERDA-E(4918)-2322.

Bennington, G.E. and Rebibo, K.K. 1976. *SPURR—System for projecting utilization of renewable resources.* McLean, Va.: The Metrek Division of the Mitre Corporation.

Benore, C.A. 1978. Dividend policy and command share valuation of electric utilities. *Edison Electric Institute financial conference.*

Berlin, E.; Cicchetti, C.J.; and Gillen, W.J. 1974. *Perspective on power.* Cambridge, Mass.: Ballinger.

Bezdek, R., editor. 1977. *Analysis of policy options for accelerating commercialization of solar heating and cooling systems.* Programs of Policy Studies in Science and Technology. Washington: George Washington University.

Bezdek, R. 1978. An analysis of the current economic feasibility of solar water and space conditioning. Washington: U.S. Department of Energy. DOE/CS-0023, UC-59.

Bezdek, R.H., Hirshberg, A.S.; and Babcock, W.H. 1979. Economic feasibility of solar water and space heating. *Science* 230:1214-1220.

Boiteaux, M. 1956. La vente au cout marginal. *Revue Francaise de l'Energie.*

Bonbright, J.C. 1961. *Principles of public utility rates.* New York: Columbia University Press.

Booz, Allen and Hamilton, Inc. 1975. Solar energy utilization in Florida. Prepared for Florida Energy Committee.

———. 1979. *Passive and active solar heating analysis in support of building energy performance standards development.* Bethesda, Maryland: Booz, Allen and Hamilton.

Boyd, D.M.; Caskey, J.F.; Price, G.A.; and Spewak, C.P. 1978. *Solar impact on gas utilities.* McLean, Va.: Mitre Corp.

Brandemuehl, M.J., and Beckman, W.A. 1978. Sensitivity analysis of solar heating systems economics. *Proceedings of the American section of the International Solar Energy Society.* Denver, Colo.

Bright, R., and Davitian, H. 1978. *Electric utilities and residential solar heating and hot water systems.* Upton, N.Y.: Brookhaven National Laboratory, EY-76-C-02-0016.

———. and ———. 1979. *The marginal cost of electricity used as a backup for solar hot water systems: A case study.* Upton, N.Y.: Brookhaven National Laboratory, BNL-25501.

Buchberg, H., and Roulet, J.R. 1968. Simulation and optimization of solar collection and storage for house heating. *Solar Energy* 12 (1):31-50.

Butz, L.W.; Beckman, W.A.; and Duffie, J.A. 1974. Simulation of a solar heating and cooling system. *Solar Energy* 16:129-136.

California Pollution Control Financing Authority (CPCFA). 1979. *Total bonds sold.* Sacramento, Calif.: CPCFA.

California PUC Energy Conservation Team. 1977. Study of the viability and cost effectiveness of solar energy application for essential uses in the residential sector of California. Sacramento, Calif.

Cargill, T.P. and Meyer, R.A. 1971. Estimating the demand for electricity by time of day. *Applied Economics* 3:233–246.

Chapman, D. 1979. Decommissioning, taxation and nuclear power costs. Testimony before the California State Assembly, August 14, 1979.

Chen, G.K.C., and Winters, P.R. 1966. Forecasting peak demand for an electric utility with a hybrid exponential model. *Management Science* 12 (12).

Cherry, B.H. 1979. *Electric load forecasting: Probing the issues with models.* EMF Report 3, vol. 1. Stanford, Calif.: Stanford University, Energy Modeling Forum.

Cicchetti, C.; Gillen, W.; and Smolensky, P. 1976. *Marginal cost and pricing of electricity: An applied approach.* National Technical Information Service, PB 255 967, for NSF/RA–760183. (Also published by Cambridge, Mass.: Ballinger Books in 1977.)

Cicchetti, C., and Jureitz, J. 1975. *Studies in electric utility regulation.* Cambridge, Mass.: Ballinger.

Claridge, D. 1977. Window management and energy savings. *Energy and Buildings* 1:57–63.

Close, D.J. 1967. A design approach for solar processes. *Solar Energy* 11:112–122.

Commoner, B. 1976. *The poverty of power, energy and the economic crisis.* New York: Knopf.

Cone, B.W., et al. 1978. (rev. ed.). *An analysis of federal incentives used to stimulate energy production.* Richland, Wash.: Battelle, Pacific Northwest Laboratory. PNL–2410–REV.

Cook, E. 1976. *Man, energy, society.* San Francisco: Freeman.

Cooper, P.I.; Klein, S.A.; and Dixon, C.W. 1975. Experimental and simulated performance of closed loop solar water heating system. Paper presented at the *International Solar Energy Congress and Exposition,* University of California at Los Angeles, July 28–August 1, 1975.

Corpening, S.L.; Reppen, N.D.; and Ringlee, R.J. 1973. Experience with weather sensitive load models for short- and long-term forecasting. *IEEE Trans.* (Poser Apparatus and Systems), PAS–92:1966–1972.

Costello, D. 1976. *Midwest Research Institute programs dealing with incentive and barriers to the commercialization of solar energy.* Working paper. Kansas City, Kan.: Midwest Research Institute.

Czahar, R. 1979. *A study of solar financing.* Sacramento: California Public Utility Commission (CPUC).

A dark future for utilities. *Business Week,* May 28, 1979, pp. 108–124.

Davidson, S. 1958. Accelerated depreciation and the allocation of income taxes. *The Accounting Review* v.33:173–180.

———. 1979. Testimony before the State of Illinois. Illinois Commerce Commission Docket No. 70–0071.

Davis, E.S., 1976. Commercializing solar energy: The case for gas utility ownership. Mimeo. Pasadena, Calif.: Jet Propulsion Laboratory, California Institute of Technology.

Davis, E.S., and Bartera, R.E. 1976. *Design and evaluation of solar-assisted gas energy water heating systems for new apartments: SAGE phase II report.* Technical Memorandum No. 5030–15. Pasadena, Calif.: Jet Propulsion Laboratory, California Institute of Technology.

Davis, E.S., French, R.L.; and Hirshberg, A.S. 1976. *Scenarios for the utilization of solar energy in southern California buildings.* Technical Memorandum No. 5040–10. Pasadena, Calif.: Jet Propulsion Laboratory, California Institute of Technology.

Davis, E.S., and Wen, L.C. 1975. *Solar heating and cooling systems for buildings: Technology and selected case studies.* Technical Memorandum No. 5040–9. Pasadena, Calif.: Jet Propulsion Laboratory, California Institute of Technology.

Debs, A. 1979. *Management of electric back-up demand for solar heating and cooling applications.* Draft. Atlanta: Georgia Institute of Technology.

Department of Energy. 1978. *National program plan for research and development in solar heating and cooling for building, agricultural, and industrial applications.* U.S. DOE/CS–0008 UC–59C. Springfield, Va.: National Technical Information Service.

Dickson, C.; Eichen, M.; and Feldman, S. 1977. Solar energy and U.S. public utilities. *Energy Policy.* vol. 5:3.

Dubin, Fred S. 1977. Energy conservation studies. *Energy and Buildings* 1:31–42.

Duffie, J.A., and Beckman, W.A. 1974. *Solar energy thermal processes.* New York: Wiley-Interscience.

Duffie, J.A., Beckman, W.A.; and Brandemuehl, M.J. 1976. A parametric study of critical fuel costs for solar heating in North America. *Proceedings of the Annual Meeting of the International Solar Energy Society,* Winnipeg, Canada, vol. 9.

Edenburn, M.W., and Grandjean, N.R. 1975. Energy system simulation computer program—SOLSYS. Paper No. SAND 75–0048. Los Alamos, N.M.: Sandia Laboratories.

Edison Electric Institute (EEI). 1977. *Statistical yearbook of the electric utility industry.* New York: EEI.

Elgerd, O.I. 1971. *Electric energy systems theory: An introduction.* New York: McGraw-Hill.

Energy Fact Book. 1976. *1976 Energy Fact Book.* Arlington, Va.: Tetra Tech, Inc.

Energy Policy Project of the Ford Foundation. 1974. *A time to choose: America's energy future.* Cambridge, Mass.: Ballinger.

EPRI. 1978a. Technical assessment guide. Palo Alto. Calif. Electrical Power Research Institute (EPRI). EPRI PS–866–SR.

———. 1978b. *Costs and benefits of over/under capacity in electric power system planning—Final report* (RP1107). Palo Alto, Calif. Decision Focus, Inc.

Fassbender, A.W., and Cone, B.W. 1978. A summary of the analysis of federal incentives used to simulate energy production. In *Proceeding of the 1978 Annual Meeting of the American Section of ISES,* Denver, pp. 400–407.

FCC Reports. 1968. 2d series, 13 (1968):420.

———. 1971. 2d series, 28 (1971):267–308.

Federal Energy Administration. 1974. *Project independence blueprint. Residential and commercial energy use patterns 1970–1990,* vol. 1.

———. 1977. *Electric utility rate design proposals.* Springfield, Va.: National Technical Information Service, FEA/D–77/063.

Federal Trade Commission, 1978. *The solar market: Proceedings.* Washington: FTC.

Feldman, S.L., and Breese, J. 1978. The economic impact of methane generation on dairy farms—A micro-analytic model. *Resource Recovery and Conservation* 3:251–273.

Feldman, S.L. 1978. Utility rates and solar energy. For the White House Domestic Policy Review of Solar Energy.

Feldman, S.L., et al. 1979. *The impact of solar energy in buildings upon California electric utilities.* Sacramento, Calif.: State of California, Energy Resources Conservation and Development Commission.

Feldman, S.L., and Anderson, B. 1975a. Financial incentives for the adoption of solar energy design. *Solar Energy* 17:339–343.

———, and ———. 1975b. Non-conventional incentives for the adoption of solar energy design. Report NSF/RANN APR–75–18006, Interim Report #2. Worcester, Mass.: Clark University.

———, and ———. 1976a. Utility pricing and solar energy design. Report NSF/RANN APR–75–18006. Worcester, Mass.: Clark University.

———, and ———. 1976b. The public utility and solar energy interface: An assessment of policy options. Report to ERDA E(49–18)–2523. Worcester, Mass.: Clark University.

———, and ———. 1977. The impact of active and passive solar building designs on utility peak loads. Interim report, DOE contract EG–77–G–01–4029. Worcester, Mass.: Clark University.

Feldman, S.L., Wirtshafter, R.M.; Abrash, M.; Anderson, B.; Sullivan, P.; and Kohler, J. 1979. The impact of federal tax policy and electric utility rate schedules upon the solar building/electric utility interface. Contract EG–77–G–01–4029, US DOE. Worcester, Mass.: Clark University.

——, and Berz, M. 1979. Barriers and incentives to the commercialization of solar energy. In P. Chereminsoff and W. Dickinson, *The Solar Energy Handbook.* New York: Marcel Dekkar.

Feldman, S., et al. 1978. The impact of solar energy upon the Texas Utilities Company. Environmetrics Corporation Report #128. Dallas, Texas: The Texas Utilities Co.

Ferguson, C.E., and Gould, J.P. 1966. *Microeconomic theory.* Homewood, Ill.: Richard D. Lewin, Inc.

Feuerstein, R.J. 1979. Utility rates and solar commercialization. *Solar Law Reporter* 1 (2):334–368.

Finder, Alan E. 1977. *The states and electricity utility regulation.* Lexington, Ky.: The Council of State Governments.

Fisher, F.M., and Kaysen, C. 1962. *A study in econometrics: The demand for electricity in the United States.* Amsterdam: North-Holland Publishing Co.

Fitch, James Marston. 1975. *American building: The environmental forces that shape it,* 2d ed. New York: Schocken Books.

Fitzpatrick, D.B., and Groebner, D.F. 1978. *The regulation of non-utility operations of public utilities in the state of Idaho.* Boise, Idaho: Boise State University.

Fitzpatrick, D.B. and Stitzel, T.E. 1978. Capitalizing an allowance for funds used during construction: The impact on earnings quality. *Public Utilities Fortnightly* 101 (2).

Flaim, T.; Considine, T.; Witholder, R.; and Edesess, M. 1979. Economic assessment of intermittent, grid-connected solar electric technologies: A review of methods. SERI/TR–353–474. Draft report. Golden, Co.: Solar Energy Research Institute.

Ford, A., and Yabroff, I.W. 1978. Defending against uncertainty in the electric utility industry, LA–UR–78–3228, LASL. Los Alamos, N.M.: Los Alamos Scientific Laboratory.

FPC. 1971. Methodology of load forecasting, Part IV of *The 1970 national power survey.* Federal Power Commission (FPC).

General Electric Corp. 1974. *Solar heating and cooling of buildings, phase O feasibility and planning study, final report.* 3 vols. Valley Forge, Pa.

Glaser, P.E., and Platte, M.D. 1975. Solar climate control, storage, and electric utility load management. *Proceedings of the Conference on the Challenge of Load Management.* Office of Utilities Programs. Washington, D.C.: Federal Energy Administration.

Graven, R.M., 1974. A comparison of computer programs used for modeling solar heating and air conditioning systems for buildings. Berkeley: University of California, Lawrence Berkeley Laboratory.

Graven, R.M., and Hirsch, P.R. 1977. *Cal-ERDA users manual version 1.3.* Argonne, Ill.: Argonne National Laboratory, ANL/ENG-77-03.

Grot, R.A., and Socolow, R.H. 1974. Energy utilization in a residential community. In *Energy,* ed. M. Macrakis. Cambridge, Mass.: M.I.T. Press.

Gupta, P.C. 1969. *Statistical and stochastic techniques for peak power demand forecasting in electric utility systems.* PEREC Report No. 51. Lafayette, Ind.: Engineering Experiment Station, Purdue University.

Gutierrez, G.; Hincapie, F.; Duffie, J.A.; and Beckman, W.A. 1976. Simulation of auxiliary energy supply, load type, and storage capacity. *Solar Energy* 15:287-298.

Habicht, E.R. 1977. Electric utilities and solar energy: Competition, subsidies, ownership and prices. Presented at the FTC Solar Symposium, Washington. Published in FTC (1978).

Halbraken, N.J., and Johnson, T.E. 1978. *Design methodologies for energy conservation and passive heating of buildings utilizing improved building components.* Progress report No. 3, Cambridge: Massachusetts Institute of Technology, C00/45133US DOE.

Halvorsen, R. 1976. Demand for electric energy in the United States. *Southern Economic Journal* 42 (4):610-625.

Hammond, Allen, and Metz, W. 1977. Solar energy research: Making solar after the nuclear model. *Science* 197:241-244.

Hastings, S.R., and Crenshaw, R.W. 1977. Window design strategies to conserve energy. *National Bureau of Standards Building Science Series 104.* Washington.

Hay, Harold. 1971. New roofs for hot dry regions. *Ekistics* 31:158-164.

Heinemann, G.T.; Nordman, D.A.; and Plant, E.C. 1966. The relationship between summer weather and summer loads—A regression analysis. *IEEE Trans.* (Power Apparatus and Systems), PAS-85:1144-1151.

Heldenbrand, J.M. 1976. *Design and evaluation criteria for energy conservation in new buildings.* Washington, D.C.: Center for Building Technology, Institute for Applied Technology, National Bureau of Standards.

Heldt, R.W. 1978. Applications of computer modeling in passive solar design. In *Proceedings of the 1978 Annual Meeting of the American Section of the International Solar Energy Society,* Denver, Colo.

Hirshberg, Alan, and Davis, E.S. 1977. *Solar energy in buildings: Implications for California energy policy.* For State of California, ERCDC.

Pasadena, Calif.: Jet Propulsion Laboratory, California Institute of Technology.

Hirst, Eric. 1979. Reducing residential energy growth. *ASHRAE Journal* 44–46.

Hisper. 1978. Synopsis of Hisper. Presented at Solar Heating and Cooling System Simulation and Economic Analysis Working Group, November 16–17, 1978. Denver, Colo.

Hittle, D.C. 1977. *The building loads analysis and system thermodynamics program, user manual.* Champaign, Ill.: U.S. Army Construction Engineering Research Laboratory. NTIS AD/A-048734.

Holt, D.D. 1979. The nuke that became a political weapon. *Fortune* 99 (1).

Houthakker, H.S., and Taylor, L.D. 1970. *Consumer demand in the United States.* 2d ed. Cambridge, Mass.: Harvard University Press.

Hull, D.E., and Giellis, R.T. 1978. Solocost: A solar energy design program. In *Conference on Systems Simulation and Economic Analysis for Solar Heating and Cooling.* SAND 78-1927. Los Alamos, New Mexico.

Huse, D., et al. 1976. The application of energy storage systems to electric utilities in the U.S. *International Conference on Large High Voltage Electric Systems.*

ICF Inc. 1977. Preliminary analysis of conservation investments as a gas utility supply option. FEA/G-77/010, January 7, 1977, Washington: ICF, Inc.

———. 1978. *Technical, institutional and economic analysis of alternative electric rate design and related regulatory issues in support of DOE utility conservation programs and policy.* Vol. 3, *Economic analysis.* Washington: ICF, Inc.

Internal Revenue Service. 1978. Computation of Investment Credit, Form 3468.

Johnson, J.R. 1979. Construction work in progress: Planning for the rate case. *Public Utilities Fortnightly* 104 (3).

Jones, J.R., and O'Donnell, J.L. 1978. Double-leverage: Lawful and based on sound economics? *Public Utilities Fortnightly* 101 (12): 26–31.

Joskow, Paul L. 1974. Inflation and environmental concern: Structural change in the process of public utility price regulation. *Journal of Law and Economics* 17 (2):291–327.

Kahn, Alfred E. 1970. *The economics of regulation principles and institutions.* Vol. 1, *Economic principles;* vol. 2, *Institutional issues.* New York: Wiley.

Kahn, E., 1979a. The compatibility of wind and solar technology with conventional energy systems. *Annual Review of Energy* 4 (1979).

———. 1979b. Project lead times and demand uncertainty: Implications for the financial risk of electric utilities. *E.F. Hutton Fixed Income Research Conference on Electric Utilities.*

Kahn, E. and Schutz, S. 1978. Utility investment in on-site solar: Risk and return analysis for capitalization and financing. LBL–7876. Berkeley, Calif.: Lawrence Berkeley Laboratory.

Kelley, H., et al. 1977. Application of solar technology to today's energy needs. Office of Technology Assessment.

Klein, S.A. 1976. A design procedure for solar heating systems. Ph.D. dissertation. Department of Chemical Engineering, University of Wisconsin-Madison.

Klein, S.A., Beckman, W.A.; and Duffie, J.A. 1976a. A design procedure for solar heating systems. *Solar Energy* 18 (2).

————; ————; and ————. 1976b. TRNSYS—A transient simulation program. *ASHRAE Transactions* 82, pt. 1.

————; ————; and ————. 1975. A method of simulation of solar processes and its application. *Solar Energy* 17 (1):29–37.

————; ————; and ————. 1973. *TRNSYS—A transient simulation program user's manual.* Engineering Experiment Station Report No. 38. Solar Energy Laboratory, University of Wisconsin-Madison.

Knasel, T.M.; Kennish, W.; Cassel, M.; and Duvall, M. 1978. *Validation methodology—status report.* Presented at Solar Heating and Cooling System Simulation and Economic Analysis Working Group, Fourth Meeting, McLean, Va. McLean, Va.: Science Applications, Inc.

Kohler, J., et al. 1977. A comparison of currently available solar energy system simulation programs. Presented at New England Solar Energy Association Conference, Hartford, Conn.

Kohler, J., and Putnam, B. 1978. Passive cost and performance comparisons. In *Passive solar state of the art, Proceedings of the 2d National Passive Solar Conference, Philadelphia, Pa.* ed. D. Prowler. Killeen, Tx.: American Section, International Solar Energy Society.

Kohler, J., and Sullivan, P. 1978. *TEANET: A numerical thermal network algorithm for simulating the performance of passive systems on a TI–59 programmable calculator.* Harrisville, N.H.: Total Environmental Action, Inc.

Kraemer, S.F. 1978. *Solar law.* New York: McGraw-Hill.

Kusuda, Tamami. 1976. *NBSLD, the computer program for heating and cooling loads in buildings.* National Bureau of Standards, Building Science Series 69. Washington, D.C.

Laitos, J., and Feuerstein, R.J. 1979. Regulated utilities and solar energy. SERI/TR–62–255. Golden, Colo. Solar Energy Research Institute.

LASL. 1978. LASL solar. Presented at the Solar Heating and Cooling System Simulation and Economic Analysis Working Group, Fourth Meeting. Denver, Colorado.

Letter from C.P. Davenport, Pacific Power and Light, to C.J. Blumstein, Lawrence Berkeley Laboratory, August 31, 1979.

Linhart, P. 1970. Some analytical results on tax depreciation. *Bell Journal of Economics and Management Science* 1 (1):82–112.

Liu, B.Y.H., and Jordan, R.C. 1963. The long-term average performance of flat-plate solar energy collectors. *Solar Energy* 7 (2):53–74.

Llavina, R. 1976. *The impact of solar heating and cooling on electric utilities.* San Juan, P.R.: Puerto Rico Water Resources Authority.

Löf, G.O.G., and Tybout, R.A. 1973a. The design and cost of optimized systems for residential heating and cooling by solar energy. Presented at the International Congress, The Sun in the Service of Mankind.

————, and ————. 1973b. Cost of house heating with solar energy. *Solar Energy* 19:253–278.

———— and ————. 1974. The design and cost of optimized systems for residential heating and cooling by solar energy. *Solar Energy* 16 (1).

Lorsch, Harold. 1977. Implications of residential solar space conditioning on electric utilities, final report. NSF-C1033 AER-75-14270. Philadelphia, Pa.: Franklin Institute.

Los Alamos Scientific Laboratory. 1976. *Passive solar heating and cooling conference and workshop proceedings.* Albuquerque, N.M.

Macrakis, Michael S. 1974. *Energy: Demand, conservation, and institutional programs.* Cambridge, Mass.: M.I.T. Press.

Manning, Jr., Willard G.; Mitchell, Bridger M.; and Acton, Jan Paul. 1976. Design of the Los Angeles peak-load pricing experiment for electricity. Report R-1955-DWP, Rand Corporation, Santa Monica, Calif.

Marx, T.G. 1978. Parents and subsidiaries and embedded cost of debt. *Public Utilities Fortnightly* 101 (11):16–19.

Mass Design. 1976. *Solar heated houses for New England and other north temperate climates.* Cambridge, Mass.: Mass Design Architects and Planners, Inc.

Maybaum, M.W. 1978. A test problem and solutions for solar heating and cooling computer simulation programs. *Proceedings of the System Simulation Conference* at Denver, Colorado. Washington: Department of Energy.

McCormick, P.O. 1975. Modification to the Lockheed-Huntsville solar heating and cooling systems simulation program, Lockheed Missiles and Space Co. Report NSF/RANN/SE/C-898/FR/75/2.

McFarland, R.D. 1978. PASOLE: A general simulation program for passive solar energy. Los Alamos, N.M.: Los Alamos Scientific Laboratory, LA 7433 MS.

Means. 1976. *Means Construction Cost Data.* Duxbury, Mass.: Robert Snow Means Co. Inc.

Meinel, Aden B., and Meinel, Marjorie P. 1976. *Applied solar energy: An introduction.* Reading, Mass.: Addison-Wesley.

Merrow, E.; Chapel, S.; and Worthing, C. 1979. *A review of cost estima-*

tion in new technologies: Implications for energy procession plants.
Santa Monica, Calif.: Rand Corporation, R–2481–DOE.

Milbank, Neil O. 1977. Energy savings and peak power reduction through the utilization of natural ventilation. *Energy and Buildings* 1:85–88.

Miller, A.S., and Thompson, G.P. 1977. *Legal barriers to solar heating and cooling of buildings.* Washington, D.C.: Environmental Law Institute for ERDA–Ex–76–C–01–2528.

Mills, G. 1977. Demand electric rates: A new problem and challenge for solar heating. *ASHRAE Journal* 42–44.

Mitalas, G.P., and Stephenson, D.G. 1967. Thermal response factors. *ASHRAE Transactions* 43, pt. 1.

Mitchell, Bridger M.; Manning, Jr., Willard G.; and Acton, Jan Paul. 1977. Electricity pricing and load management: Foreign experience and California opportunities. Report R–2106–CERDC. Santa Monica, Calif.: Rand Corporation.

Montgomery, W. David. 1977. A case study of regulatory programs of the federal energy administration. In *Study on Federal Regulation,* vol. 6. Committee on Government Operations, United States Senate. Washington: Government Printing Office.

Moody's Public Utility Manual. 1978. New York: Moody's Investor Services Inc.

Morgan, M.G., and Talukdar, S.N. 1979. Electric power load management: Some technical, economic, regulatory and social issues. *Proceedings of the IEEE* 67 (2):241–312.

Morgan, Richard. 1976. *Phantom taxes in your electric bill.* Washington: Environmental Action Foundation.

Mossin, J. 1973. *The theory of financial markets.* Englewood Cliffs, N.J.: Prentice-Hall.

Mount, T.D.; Chapman, L.D.; and Tyrrell, T.J. 1973. Electricity demand in the United States: An econometric analysis. Oak Ridge, Tenn.: Oak Ridge National Laboratory (ORNL–NSF–49).

Mulligan, G. Legislative history of the investment tax credit. Undated CPUC memo.

Murray, H., et al. 1974. Control systems design and simulation for solar heated structures. Los Alamos, N.M.: Los Alamos Scientific Laboratory, LA–UR–74–1085.

Mutch, J.J. 1974. Residential water heating, fuel consumption, economics and public policy. Report R1498. Los Angeles: Rand Corporation.

Myers, S.C.; Dill, D.A.; and Bautista, A.J. 1976. Valuation of financial lease contracts. *Journal of Finance* 31:799–819.

Nelson, J.R. 1964. *Marginal cost pricing in practice.* Englewood Cliffs, N.J.: Prentice-Hall.

Noll, Roger G., and Rivlin, Lewis A. 1973. Regulating prices in competitive markets. *Yale Law Journal* 82 (7):1426–1434.

Northeast Utilities. 1977. *Connecticut peak load pricing test.* Hartford, Conn.: Northeast Utilities Company.

Odum, H.T., and Odum, E.C. 1976. *Energy basis of man and nature.* New York: McGraw-Hill.

Olgyay, Victor. 1963. *Design with climate, bioclimatic approach to architectural regionalism.* Princeton, N.J.: Princeton University Press.

Olson, C.E. 1975. Public utility regulation and its impact on electricity demand and production. In *Studies in electric utility regulation,* eds. C. Cicchetti and J. Jurewitz. Cambridge, Mass.: Ballinger.

Oonk, R.L.; Beckman, W.A.; and Duffie, J.A. 1975. Modeling of the CSU heating/cooling system. *Solar Energy* 17:21–28.

Opinion Research Corp. 1976. *Barriers to energy conservation.* FEA Conservation Paper No. 55. Washington, D.C.

O'Riordan, Timothy. 1976. *Environmentalism.* London: Pion Ltd.

O'Toole, James R. 1976. *Energy and social change.* Cambridge, Mass.: M.I.T. Press.

Passive Solar Heating and Cooling Conference. 1976. *Passive solar heating and cooling.* Washington: Energy Research and Development Administration.

Peles, Y.C., and Stein, J.L. 1976. The effect of rate return regulation is highly sensitive to the nature of the uncertainty. *American Economic Review* 66 (1976):278–289.

People and Energy. 1976. Utilities and solar energy: Will they own the sun? Washington, D.C.

Peterson, H. Craig. 1976. The impact of tax incentives and auxiliary fuel prices on the utilization rate of solar energy space conditioning. NSF/RANN Report. Springfield, Va.: National Technical Information Service.

Peterson, H. Craig, and Peterson, J.E. 1979. Estimating air infiltration into houses: An analytical approach. *ASHRAE Journal* 60–62.

Peterson, S.R. 1974. *Retrofitting existing housing for energy conservation: An economic analysis.* Building Science Series 64. Washington: Center for Building Technology, Institute for Applied Technology.

Pindyck, R.S. 1974. The regulatory implications of three alternative econometric supply models of natural gas. *Bell Journal of Economics* 5 (2):633–645.

Platts, J.E., and Womeldorff, P.J. 1979. The significance of assumptions used in long-range electric utility planning studies. IEEE Power Engineering Society Conference Paper F/79/772–5. Vancouver, British Columbia.

Pomerantz, L.S., and Suelflow, J.E. 1975. *Allowance for funds used during construction: Theory and application.* East Lansing, Mich.: Michigan State University Public Utilities Studies.

Posner, Richard A. 1971. Taxation by regulation. *Bell Journal of Economics* 2 (1):22–50.

Prentice-Hall 1975 Federal Tax Course. 1974. Englewood Cliffs, N.J.: Prentice-Hall.

Priest, A. 1969. *Principles of public utility regulation.* 2 vols. Charlottesville, Va.: Michie Co.

Roach, Fred, et al. 1977. Prospects for solar energy: The impact of the national energy plan. Los Alamos, N.M.: Los Alamos Scientific Laboratory.

Roach, F., et al. 1978. The comparative economics of passive and active systems: Residential space heating applications. In *Proceedings of the 1978 Annual Meeting of the American Section of ISES,* pp. 537–541, Denver, Colo.

Ruegg, R.T. 1976a. *Evaluating incentives for solar heating.* NBS 1R76–1127. Washington: National Bureau of Standards.

———. 1976b. *Solar heating and cooling in buildings: Methods of economic evaluations.* NBSIR75–712. Washington: National Bureau of Standards.

Rumelt, R. 1974. *Strategy, structure and economic performance.* Cambridge, Mass.: Harvard University Press.

Samuelson, P.A. 1947. *Foundations of economic analysis.* Cambridge, Mass.: Harvard University Press.

San Diego's utility typifies industry woes. *Business Week,* May 28, 1979, p. 110.

Sawyer, Stephen, and Feldman, S.L. 1978. The barriers and incentives to the commercialization of solar energy for residential domestic hot water and space heating: An in-depth assessment by solar consumers. *Proceedings of the 1978 Annual Meeting of the American Section of International Solar Energy Society,* Denver, Colo.

Schiffel, D.; Costello, D.; and Posner, D. 1978. Projecting the market penetration of solar energy: A methodological review. In *Proceedings of the 1978 Annual Meeting of the American Section of the International Solar Energy Society,* pp. 382–386, Denver, Colo.

Schmalensee, R.L. 1978. Promoting competition in tomorrow's markets for solar energy systems. *The Solar Market: Proceedings of the Symposium on Competition in the Solar Energy Industry.* Washington: Federal Trade Commission.

Schoen, Richard; Hirshberg, A.S.; and Weingart, J.M. 1975. *New energy technologies for buildings,* ed. Jane Stein. Cambridge, Mass.: Ballinger.

Schulze, W.D., and Ben David, S. 1977. *The economics of solar home heating.* Published for the Joint Economic Committee of Congress. Washington: Government Printing Office.

Science Applications, Inc. 1978. *Survey of currently used simulation methods.* McClean, Va.: Science Applications, Inc.

Seeds, K.J. 1978. The direct approach to subsidiaries' capital costs. *Public Utilities Fortnightly* 101 (10):18–20.

Sharefkin, M. 1974. *The economics and environmental benefits from improving electrical rate structures.* Chevy Chase, Md.: Jack Faucett.

Sharpe, W.F. 1970. *Portfolio theory and capital markets.* New York: McGraw-Hill.

Shelpuk, B.; Joy, P.; and Crothamel, M. 1976. *Technical and economic feasibility of thermal storage.* Camden, N.J.: PCA Advanced Technology Laboratories. ERDA contract EY–76–C–02–2591, COO/2591–76/1.

Sheridan, N.R.; Bullock, K.J.; and Duffie, J.A. 1967. Study of solar processes by analog computer. *Solar Energy* 11 (2):69–77.

Shurcliff, W.A. 1978. *Solar heated buildings of North America: 120 outstanding examples.* Church Hill, Harrisville, N.H.: Brick House Publishing Co.

Smartt, L.E. 1979. Some tax considerations for the utility industries. *Public Utilities Fortnightly* 143 (4).

Smith, R.O., and Meeker, J. 1978. New England Electric's solar project. *Solar Age* 16–24.

Snell, J.E.; Achenbach, P.R.; and Petersen, S.R. 1976. Energy conservation in new housing design. *Science* 192:1305–1311.

Socolow, R.H., et al. 1978. The Twin Rivers program on energy conservation in buildings. *Energy and Buildings* 211–230.

Solar Energy Laboratory. 1975. *TRNSYS, A transient simulation program.* Report #38. Madison, Wis.: University of Wisconsin–Madison.

Southern California Gas Company. 1976. *Project Sage: Solar assisted gas energy project.* DSE/4691–76/1. Los Angeles, Calif.

Southwest Energy Management Inc. 1978. *Multi-Family solar water heating.* San Diego, Calif.

Sparrow, F.T. 1979. *Solar-utility interface issues.* Unpublished paper. Purdue University.

Stanford Research Institute. 1977. *Solar energy in America's future: A preliminary assessment.* Washington: Energy Research Development Administration, Division of Solar Energy.

Stanton, K.N., and Gupta, P.C. 1969. Long-range forecasting of electrical demand by probabilistic methods. *Proceedings of the American Power Conference* 31:964–969.

Steadman, Philip. 1975. *Energy, environment and building.* Cambridge, England: Cambridge University Press.

Stein, R. 1977. *Architecture and energy.* Garden City, N.Y.: Anchor/Doubleday Press.

Suelflow, J. 1973. *Public utility accounting: Theory and applications.* Public Utilities Studies. East Lansing: Michigan State University.

Sussman, M.R. 1979. Capital costs, double leverage and capital structure imputation. *Public Utilities Fortnightly* 103 (3):34–35.

Swetnam, G.F., and Jardine, D.M. 1976. Energy rate initiatives study of the interface between solar and wind energy systems and public utilities. Draft, Mitre Corp. Technical Report 7431.

Systems Controls. 1977. *Critical analysis of European load management practices.* ERDA, CONS/1168–1 NTIS. Palo Alto, Calif.

Taff, D.C., et al. 1978. Active versus passive systems—a design rationale for the selection of passive systems based on a comparative study of efficiency, capacity, economics and thermal network analysis. In *Passive solar: State of the art, Proceedings of the 2nd National Passive Solar Conference,* Philadelphia, Pa., ed. D. Prowler. Killeen, Tx: American Section, International Solar Energy Society.

Taylor, L.D. 1975a. A review of load forecasting methodologies in the electric utility industry. Presented at the EPRI Workshop on Time of Day and Seasonal Load Forecasting, Pacific Grove, Calif.

———. 1975b. The demand for electricity: A survey. *Bell Journal of Economics and Management Science* 6 (1).

Telkes, M. 1974. Storage of heating and cooling. Paper presented at the Annual Meeting of ASHRAE, Montreal.

Testimony of R.L. Bertschi. 1979. Before the Illinois Commerce Commission, Docket No. 79–0071.

Testimony of Eugene Coyle. 1978. Before the New Jersey Board of Public Utilities.

Testimony of T.A. Keefe. 1979. Revenue Requirement Division, California Public Utilities Commission, CPUC–OII–42, San Francisco.

Testimony of J. Shue. 1978. Before the Public Utility Commissioner of the State of Oregon, Docket UF 344.

Testimony of D.W. Sloan. 1979. Before the Montana Public Service Commission.

Tolley, G.S.; Upton, C.W.; and Hastings, S.V. 1977. *Electric energy availability and required growth.* Cambridge, Mass.: Ballinger.

Trombe, F.; Robert, J.F.; Cabanet, M.; and Sesolis, B. 1976. Some performance characteristics of the CNRS solar house collectors. In *Passive solar heating and cooling conference.* Washington.

Trout, R. 1979. A rationale for preferring construction work in progress in the rate base. *Public Utilities Fortnightly* 103:22–26.

TRW Systems Group. 1974. *Solar heating and cooling of buildings, phase O, final report.* Redondo Beach, Calif.

Turvey, Ralph. 1968. *The optimal pricing and investment in electricity supply.* Cambridge, Mass.: M.I.T. Press.

Tybout, R.A., and Löf, G.O.G. 1970. Solar energy heating. *Natural Resources Journal* 10: 268–326.

U.S. v. Western Electric and AT&T. 1956. Consent Judgement, U.S. District Court of New Jersey, 13 RR 2143; Trade Cases 71,134.

Ward, John C.; Löf, G.O.G.; and Hadley, L.N. 1976. *Maintenance cost of solar air heating systems.* EY–76–S–02–ERDA, COO/2830–1–2830. Fort Collins, Colo.: Colorado State University.

Weidenbaum, Murray L. 1976. *Financing the electric utility industry.* New York: Edison Electric Institute, EEI Pub. No. 75–46D/5M.

Wellisz, S.H. 1963. Regulation of natural gas pipeline companies: An economic analysis. *Journal of Political Economy,* February 1963, pp. 30–43.

Wessler, E.J. 1980. The role of solar hot water heating of residential buildings in electric utility load management. Ph.D. dissertation, Clark University, Worcester, Mass.

Westinghouse Electric Corporation. 1974. *Solar heating and cooling of buildings, phase O, final report.* 4 vols. Baltimore, Md.

Whillier, A. 1953. Solar energy collection and its utilization for space heating. Ph.D. thesis, Department of Mechanical Engineering, M.I.T., Cambridge, Mass.

White, Lynn. 1962. *Medieval technology and social change.* New York: Oxford University Press.

Willey, W.R.Z. 1978. *Alternate energy systems for Pacific Gas & Electric Company: An economic analysis.* Berkeley, Calif.: Environmental Defense Fund.

Williams, R.H. (ed.) 1978. *Toward a solar civilization.* Cambridge, Mass.: M.I.T. Press.

Wilson, J.W. 1971. Residential demand for electricity. *Quarterly Review of Economics and Business* 11 (1):7–22.

Winn, C.B.; Johnson, G.R.; and Corder, T.E. 1974. SIMSHAC—A simulation program for solar heating and cooling of buildings. *Journal of Simulation* 165–174.

Winn, C.B.; Parkinson, B.W.; and Duong, N. 1978. Validation of solar systems simulation programs. In *Proceedings of the American Section of the International Solar Energy Society,* Denver, Colo.

Zimmermann, Erich W. 1964. *Introduction to world resources.* ed. Henry L. Hunker. New York: Harper & Row.

1978 Annual Report of Pacific Power & Light Co. to the California Public Utilities Commission.

About the Authors

Stephen L. Feldman is an associate professor at the School of Public and Urban Policy of the University of Pennsylvania. Previous to that position, he was resident scholar and program manager for technical policy studies at the Russell Sage Foundation in New York City. Dr. Feldman has received awards from the National Bureau of Standards for the invention of a number of energy-related patents. He has been an advisor to the National Laboratories at Brookhaven and Berkeley, the California Energy Commission, the Prime Minister's Office of the State of Israel, the Wisconsin Public Service Commission, the U.S. Department of Energy, the National Science Foundation, and several electric utilities. He has co-authored a text on urban water forecasting, *The Demand for Urban Water,* and has written numerous articles on production-function analysis in water supply.

Robert M Wirtshafter received the Ph.D. from the Graduate School of Geography at Clark University. After working for several years as a private consultant on solar-energy policy, he recently joined the Tennessee Valley Authority as a policy planner. At TVA his responsibilities include overall program management issues regarding decentralized energy production. Dr. Wirtshafter has published a number of articles on the simulation modeling of decentralized energy sources for policy purposes.